Beyond Brains: Evolution's Puzzle

Nami

I Table of Contents

1 Introduction

1.1 What made us human?

During more than 200 million years of mammalian evolution, a tremendous number of genomic changes occurred and gave rise to a great variety of species.

90-100 million years ago (mya) the lineage leading to species including the great apes separated from the rodent line (Nei et al., 2001). Within this primate lineage, old world monkeys (e.g., macaques) diverged about 23 mya gradually followed by the ancestors of the modern gibbons, orangutans, and gorillas (Chen et al., 2001; Kuhlwilm et al., 2016). Finally, the human lineage diverged about 5-7 mya from the lineage of our closest living evolutionary ancestors the chimpanzees and bonobos (Glazko & Nei, 2003).

The complexity of genomic adaptations that occurred during mammalian evolution is striking, but none led to changes affecting the life on earth comparable to the ones which made us human. Today, we know that during the first half of these 5-7 million years of human evolution our earliest ancestors did not have a higher impact on the environment than jellyfishes, butterflies or chimpanzees until this existence as a usual animal began to change dramatically about 2.5 mya when the first Homo specimens arose.

Questions about this development that made us human fascinated humankind throughout the last centuries and led to answers which identified the brain as the core component of the human identity (Sousa et al., 2017). Thus, of the various adaptations during the last 2.5 million years, the expansion of the human brain, particularly of its evolutionary youngest part, the neocortex (see: 1.2), has to be highlighted. Interestingly, the increase in brain size reached a plateau about 100.000 years ago, when modern humans successfully began to colonize the world outside of Africa (Reyes & Sherwood, 2015). Today, it is an interdisciplinary challenge to understand how the expanded brain affected the far-reaching milestones in behavior and cultural evolution, which crucially defined us as a human species. One interesting clue is given by interspecies comparative anatomical analysis of brains showing an enhanced tendency of neocorticalization (ratio of neocortical volume over the total brain volume) in the human lineage combined with a relative expansion of association cortices compared to primary cortex areas (Buckner & Krienen, 2017). While the specific causalities between brain size, neuronal number and intelligence are still controversially discussed, one of the most widespread assumptions in the field remains that neocortex size and human cognition correlate positively (Jerison, 1973; Williams & Herrup, 1988; Reader & Laland, 2002; Sousa et al., 2017). Hence, the increased neocortical size is thought to have been enormously critical for the further development of the human species. The resulting cognitive capacity even made it possible to become widely independent from evolutionary pressure, because of intellectual achievements such as powerful health care systems. After this *cognitive revolution* (Harari, 2014) the most influential changes were suddenly no longer primarily coded in the genome but in our imagination and language - what makes it extraordinary exciting to identify the genomic changes that brought us to such a state. Though even the natural curiosity of many humans might be a reason enough to study this subject, insights about the genetic differences underlying the evolution of humankind will be of even higher importance if they reveal new mechanisms about the biology of neurological and psychiatric diseases affecting modern humans.

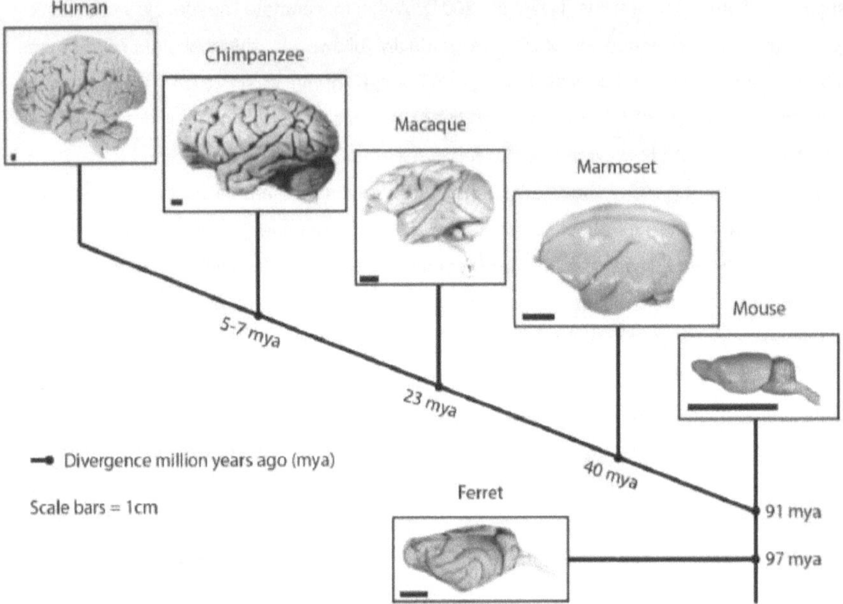

Fig. 1 – Brain expansion in the mammalian lineage

The human brain is the biggest one observed in a living species within the primate lineage. The figure shows images from an adult human, chimpanzee, macaque, marmoset, ferret and mouse brain to show the variability in size but also regarding the shape of the surface (gyrencephalic: human, chimpanzee, macaque, ferret vs. lissencephalic: marmoset, mouse) that occurred over the course of 100 million years of mammalian evolution. The shown mammalian brain images were obtained from http://www.neurosciencelibrary.org.

1.2 The Neocortex

1.2.1 Origin and structure

As the prefix suggests, the neocortex is the evolutionarily youngest part of our brain and a hallmark of mammals (Meredith et al., 2011; O'Leary et al., 2013). Early small mammals, that emerged during the Triassic/Jurassic period (about 200 mya) from their reptilian ancestors were the first species on earth having the typical uniform, six-layered sheet consisting of radially aligned neurons called the neocortex (Kaas, 2013; Florio & Huttner, 2014; Briscoe & Ragsdale, 2018). After that, the neocortex expanded independently in various mammalian lineages (Borrell & Reillo, 2012), reflecting a positive selection for an increased number of neurons. Interestingly, neocortex expansion is particularly evident in anthropoid primates and especially in humans. Two-thirds of our overall brain mass is constituted by the neocortex. (Florio & Huttner, 2014)

Simplified the neocortex could be described as a thin mantle of gray matter (GM) enclosing the underlying white matter (Rakic, 2009). In contrast to lissencephalic species, the neocortical white and grey matter of a gyrencephalic species form ridges (gyri) surrounded by furrows (sulci) on the cerebral cortex defining its specific shape (Fig. 2).

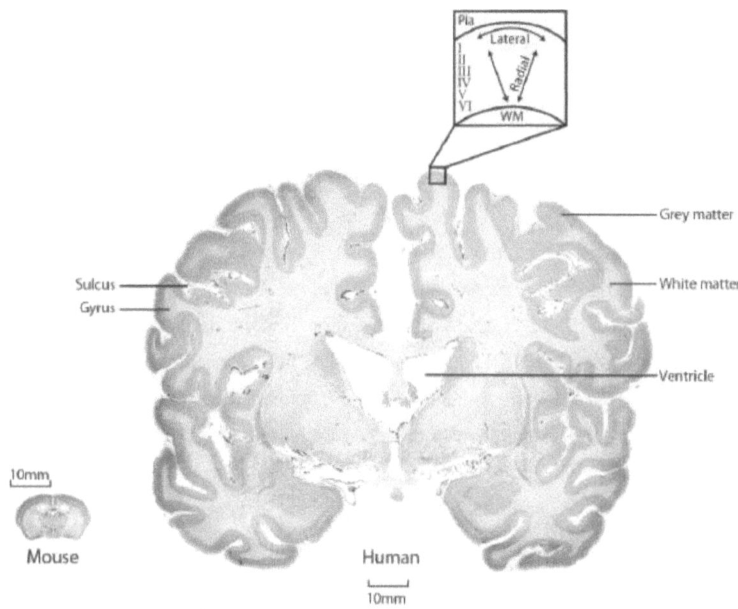

Fig. 2 – Morphology of the neocortex

Coronal sections of an adult gyrencephalic human (right) and an adult lissencephalic mouse (left) brain. The general dimensions used to describe the neocortex (i) radial: along the white matter (WM) to-pia-axis, corresponding to the apical basal axis in terms of tissue polarity; and (ii) lateral: along the axis perpendicular to the radial axis, are outlined in the inset at the top as well as the organization of the cortical layers I, II, III, IV, V & VI in the grey matter. The figure was adapted from Fig. 1 of Florio & Huttner (2014). Mouse neocortex adapted with permission from the *High Resolution Mouse Brain Atlas* (Sidman et al., 1999), http://www.hms.harvard.edu/research/brain; human neocortex adapted with permission from http://www.brains.rad.msu.edu and http://brainmuseum.org (supported by the US National Science Foundation).

The white matter mainly consists of myelinated axons stemming from or directed to the gray matter. The latter one comprises a network of excitatory ($\approx$80% of total neurons) and inhibitory ($\approx$20% of total neurons) neurons, glial cells and blood vessels. In the lateral dimension, the gray matter is arranged in six layers or laminae I-VI with I being the outermost and VI the innermost one (Fig. 2). Neurons in one layer share similar cell type identities and connectivity patterns, thus upper-layer neurons (from layer II-III), for instance mainly connect to other cortical areas. In contrast, the majority of neurons in layer IV receive input from outside the cortex, mostly from the *gateway to consciousness* – the thalamus. Neurons from layers V and VI often project to subcortical structures, e.g., to the thalamus, brainstem, and spinal cord (Noback, 2005).

Local varieties in neuron type, density and connectivity subdivide the GM into various distinct areas representing functionally specialized cortical fields. Those can be distinguished in primary and association (secondary and tertiary) areas.

1.2.2 Neurogenesis in the developing neocortex

After the edges of the neural plate appeared as neural folds during the third week of development and thereby confined the neural channel, the margins of this channel approach each other and finally merge entirely forming the neural tube.

The neural tube is made of a pseudostratified layer of neuroepithelial cells (NEC) undergoing a period of proliferative divisions which affect the growth of the neocortex in two dimensions: (1) expansion in the lateral and (2) thickening of the neuroepithelium, i.e., growth in the radial dimension (Rakic, 1995), (Florio & Huttner, 2014), before NECs begin to differentiate into different classes of neural cells. In mice, cortical neurogenesis starts at mid-gestation between E9-E10. At this time, single NECs switch to an asymmetric differentiative cell division (Götz & Huttner, 2005; Huttner & Kosodo, 2005; Martynoga et al., 2012) and transform into the apical radial glial cell (aRG) (Hartfuss et al., 2001; A. R. Kriegstein & Götz, 2003), aRGs can divide asymmetrically and produce the second progenitor cell type, basal progenitors (BPs) (Florio & Huttner, 2014). There are mainly two types of BPs: basal intermediate progenitor cells (bIPs) and basal radial glial cells (bRGs, also known as outer RGs). In mouse, most of BPs are neurogenic bIPs that divide symmetrically and directly produce neurons. In contrast, in primates and most in humans, there are substantial numbers of bRGs. Because bRGs are highly proliferative and finally generate many neurons, this progenitor cell type is thought to be crucial for neocortical expansion. The occurrence of a variety of cell types in the developing neocortex finally leads to the formation of several zones. From the apical to the basal side of the neocortex, there are mainly four zones: ventricular zone (VZ), subventricular zone (SVZ), intermediate zone (IZ) and cortical plate (CP). VZ and SVZ constitute the germinal zones (GZs), which mainly consist of progenitor cells and newborn neurons. IZ is defined by the existence of axons. The most abundant cell type in this zone are migrating neurons. CP consists of neurons which are going to constitute the gray matter in the adult brain (Florio & Huttner, 2014; T. Namba & Huttner, 2017).

In the following sections, the recent knowledge of neocortical progenitor cells will be summarized.

1.2.2.1 Classes of neural progenitor cells (NPCs)

In the mammalian neocortex, there are mainly two types of NPCs. The first ones are apical progenitor cells (APs) that are inserted into the apical junctional belt and undergo mitosis at the apical surface. The second type are basal progenitor cells (BPs) that are delaminated from the apical junctional belt and therefore undergo mitosis in the basal side, often in the SVZ. Because most of the neurons are derived from BPs, the subtypes and their abundance are the critical determinant of the neuronal number and thus important for brain expansion in evolution.

1.2.2.1.1 Apical progenitors

As introduced above NECs are the first population of neural stem cells in the developing brain constituting the early neural tube. As a consequence, all neurons in the mammalian neocortex are descendants from these cells.

Polarized along the apical-basal axis, they span the entire width of the neuroepithelium contacting the overlying basal lamina with their basal plasma membrane and facing the lumen of the neural tube, the future ventricle, with the apical plasma membrane (Huttner & Brand, 1997; Götz & Huttner, 2005). Neighboring NECs are linked through a belt of Adherens junctions (AJ) at the apical-most end of their lateral membrane (Aaku-Saraste et al., 1996; Chenn et al., 1998; Marthiens & ffrench-Constant, 2009).

With the onset of cortical neurogenesis, a second type of NPCs is generated in asymmetric differentiative cell divisions – the apical radial glial cell (aRG) (Götz & Huttner, 2005; Huttner & Kosodo, 2005).

While thickening of the cortical wall, aRGs remain their contact to the basal lamina. For that reason, their basal part has to elongate and to transform into a long, thin radial fiber – the basal process, which creates a scaffold for migrating neurons (Rakic, 1972). aRGs express astroglial markers (Malatesta et al., 2000; Campbell & Götz, 2002) and upregulate certain transcription factors, notably Pax6 (Götz et al., 1998; Warren et al., 1999; Estivill-Torrus et al., 2002; Osumi et al., 2008) and Sox2 (Suh et al., 2007; Kriegstein & Alvarez-Buylla, 2009; Hansen et al., 2010; Gertz et al., 2014; Sun & Hevner, 2014; Pollen et al., 2015). Sox2 is used as a radial glial cell marker in this book.

aRGs remain highly related to NEC sharing many characteristics like the integration into the AJ belt and their apical-basal polarity, contacting both the ventricle and the basal lamina (Götz

& Huttner, 2005). They undergo either symmetric proliferative or asymmetric differentiative cell divisions, albeit proliferative divisions prevail in NECs (Florio & Huttner, 2014).

1.2.2.1.2 Basal progenitors

In the case of basal progenitors (BPs), two major cell types can be distinguished: basal radial glial cells (bRGs) and basal intermediate progenitors (bIPs) – both are directly or indirectly generated by APs.

Basal intermediate progenitors are non-epithelial cells since they delaminate from the AJ belt and do not show an apical-basal polarity anymore. Astroglial markers are downregulated in bIPs, in contrast, Tbr2 (T-box brain protein 2, EOMES) a marker for the neurogenic NPCs, is upregulated (Englund et al., 2005; Kowalczyk et al., 2009).

Within the bIPs, two subtypes were identified – neurogenic bIPs, which can be characterized as self-consuming and neurogenic (Haubensak et al., 2004; Noctor et al., 2004) leading to a doubling of the neuronal output. The other subtype is a proliferative bIP, which is capable of undergoing one more round of symmetric-proliferative divisions before a self-consuming final division (Noctor et al., 2004; Hansen et al., 2010).

Basal radial glial cells were initially characterized in ferrets and humans as monopolar BPs forming a basal process that contacts the basal lamina, but no apical process towards the ventricle (Fietz et al., 2010; Hansen et al., 2010; Reillo et al., 2011). Now, two additional subtypes were identified: bipolar bRGs, called "bRG-both-P" with processes towards the apical and basal surface and monopolar bRGs which lack a basal but exhibit an apical process that does not reach the ventricle called "bRG-apical-P" (Betizeau et al., 2013; Pilz et al., 2013; Stahl et al., 2013). Independently from the subtype, the majority of bRGs expresses Pax6 and Sox2. In contrast to aRGs, nearly half of the bRGs were found to co-express Tbr2 during interphase in the developing macaque neocortex (Betizeau et al., 2013). Regarding the evolutionary expansion of the neocortex, bRGs are of particular interest because of their capacity to self-amplify and thereby to increase the progenitor cell pool massively, which finally gives rise to neurons.

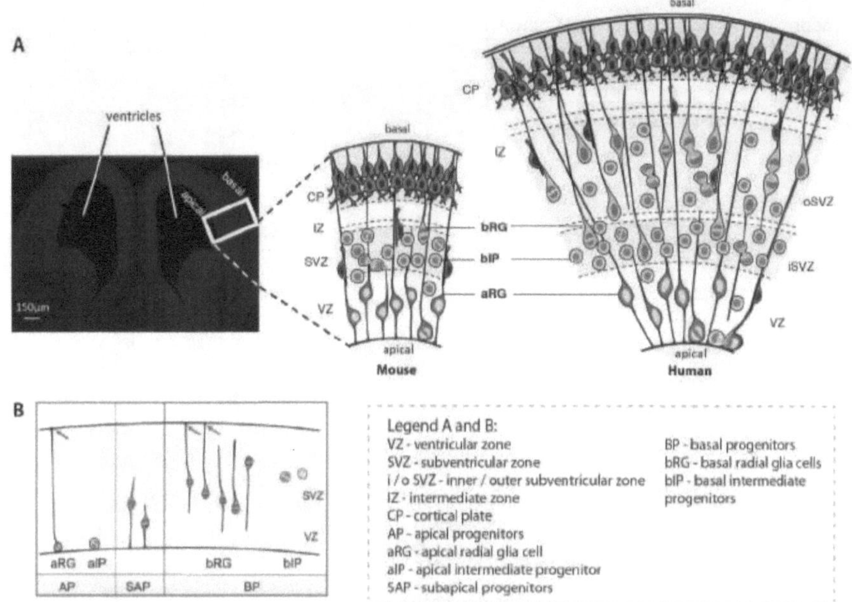

Fig. 3 – Germinal zones and NPC types in the developing neocortex

(A) Coronal section of a mouse brain at E 14.5 stained with DAPI (blue, cell nuclei) to illustrate the apical and basal orientation relative to the ventricles and the pial surface (left) and a schematic magnification of the white rectangle and of its analogon in humans. The magnification illustrates the different germinal zones, the relative abundance of the residing progenitor cell types as well as the neurons migrating along the basal processes of radial glial cells towards the cortical plate (CP).

(B) Schematic NPC types in the mammalian neocortex systematically classified according to cell polarity, presence of ventricular contact and the mitotic location. Contact to of the basal process with the basal lamina is indicated by red arrows. Neuroepithelial cells like the primary stem cells are not represented on this image. The figure was adapted from Fig. 2 of Florio & Huttner (2014).

1.2.2.2 NPC division modes

Besides, the modes of NPC cell division are also critical for brain expansion. There are two modes of cell division: symmetric and asymmetric. Symmetric cell division is further subdivided into differentiative and proliferative cell divisions.

A cell division that leads to two identical daughter cell types, which can also be different from the mother cell, is called symmetric cell division. If one of the two daughters differ from each other, it becomes asymmetric, e.g., like at the onset of neurogenesis, when the division of a NEC gives rise to an aRG and IP.

To be a proliferative cell division, the two daughter cells have to adopt the identity of the mother. If at least one daughter withdrawals from cell cycle as a neuron, the division is called neurogenic. Hence, self-amplifying, self-renewing, and self-consuming divisions can be distinguished and observed throughout neural development. Further variants of possible modes of cell division are nicely shown in Fig. 4.

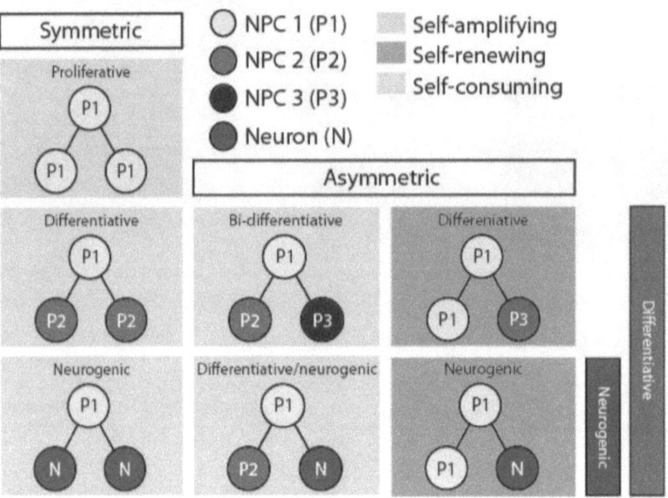

Fig. 4 – Division modes of neural progenitor cells

Different modes of NPC divisions are distinguished based on the identity of the two daughter cells relative to each other, to the parent cell and to the overall cell pool (self-amplifying, self-consuming or self-renewing). Thus, a cell division can be symmetric or asymmetric and proliferative, differentiative or neurogenic. The figure is adapted from Box 1 of Florio & Huttner (2014).

1.2.3   How to increase the neuronal output

The human neocortex is about 1000 times larger than the mouse and 2-3 times larger compared to the chimpanzee neocortex (Fig. 1) – an increase mainly based on about twice as many neurons (Striedter, 2005; Rakic, 2009). To study the underlying mechanisms, it is essential to identify parameters that could affect the neuronal output in these species and especially humans. Previous studies have shown that the expansion of the BP pool has been essential for the neocortical enlargement in the primate lineage. As a result, two distinct zones developed within the subventricular zone (SVZ): an outer (oSVZ) and inner SVZ (iSVZ). The latter one is closer related to the SVZ found in lissencephalic rodents with a higher density of progenitor cells than found in the outer SVZ (oSVZ) which is thicker offering more space to divide detached from the constricted ventricular surface (Fig. 3) (Florio & Huttner, 2014). As introduced above, BPs differ in their self-amplification capacity. Therefore, it is not only of interest to know the absolute but also the relative numbers of different BP subtypes throughout neurogenesis. This proportionate distribution of NPC subtypes is affected by intrinsic and extrinsic factors which are nicely reviewed from Florio an Huttner (2014). One of the critical parameters in the NPC biology affecting the fate towards proliferation or neurogenesis is found in the cell cycle length and composition (Calegari et al., 2005; Lange et al., 2009; Arai et al., 2011; Betizeau et al., 2013; Kinoshita et al., 2013). But also controlled cell death is part of a normal neocortex development (Haydar et al., 1999).

Last but not least, the length of the neurogenic period is one of the critical parameters that are potentially offering NPCs a longer timeframe to proliferate and to differentiate into neurons during development (Lewitus et al. 2014).

1.3   Evolution and gene duplication

1.3.1   Gene duplication and evolutionary novelty

When Darwin published his *Origin of Species* in 1859, he gave two challenges to himself. The first was to demonstrate the broad kinship of life on earth. The underlying process he termed *descent with modification*. The second was to prove that the responsible activity for these modifications can be described as natural selection (Darwin, 1959).

There are two fundamental questions one has to answer to understand the origin of novel functions: first, "what is the genetic source of the novel structure?" and "how has that new structure become adapted to its function?" (Conant & Wolfe, 2008).

The idea of gene duplication as a 'genetic backup' to be crucial for functional innovation arose already in the early 20th century (Conant & Wolfe, 2008). In 1970, finally *Evolution by Gene Duplication* was published by Susumu Ohno and elucidated for the first time the potential of gene duplication comprehensively (Hurles, 2004), after he suggested already three years earlier, that gene duplication would be the most important evolutionary force since the emergence of the universal common ancestor (Ohno, 1967). But what makes genetic duplication concretely such a key player in evolution?

A significant aspect of this question is addressed by the genetic redundancy, which is created through the existence of a second gene copy that provides the opportunity to explore this *forbidden space of evolution*. If one of the genes experiences a mutation that affects its original function, the second copy can serve as a *spare part* (Hurles, 2004) and continues to function correctly, thus also the functional innovation of the altered gene copy is facilitated.

Generally spoken innovation needs two requirements: motive and opportunity. For the evolution of a novel gene function, the predominant purpose is the gain of a novel selective advantage – with gene duplication as a major opportunity (Hurles, 2004).

But how do we know that this is more than an interesting theoretical consideration? One of the useful hints that gene duplication matters indeed is the widespread existence of gene families. If members of a gene family shared a common ancestor and arose from a duplication event, they are termed paralogues, whereas genes in different genomes sharing a common ancestor as a result of a speciation event are called orthologues (Hurles, 2004).

Whole genome sequencing of changes in gene complements of closely related species over relatively short evolutionary distances made it possible to reveal the vast expansions and contractions of gene families that can be related to the biological differences that occurred. The sensory reliance on sight is one glaring example of such a case where humans and mice differ due to duplications of the underlying gene.

In this particular case, the duplication of the red-opsin gene leading to the green-opsin gene significantly enhanced the human color vision as it allows us to distinguish light at three different wavelengths and not only at two like mice (Nei et al., 1997; Goymer, 2007). In contrast, in mice, the proportion of the large gene family of olfactory receptors that retains its functionality is a lot higher than in humans (Young et al., 2002; Gilad et al., 2003).

Even though it became a widespread belief that gene duplication facilitates adaptation, the actual importance remained in dispute. Duplication may lead to a doubling of the gene dosage or even neofunctionalization (see: 1.3.3) increasing the fitness of an individual, but could also divide the ancestral functions into daughter genes not promoting adaptation (Force et al., 1999).

This criticism of the adaptation by gene duplication hypothesis was supported by a genome-scale experiment where pairs of duplicated genes and singleton genes were deleted from the yeast genome with similar fitness effects. They concluded that duplication would rarely result in adaptation (Dean et al., 2008). Albeit, this approach neglects a known duplication bias among genes with different fitness contributions. To elude this problem, homologous genes from two yeast species were compared. In this case, the simultaneous deletion of a duplicated gene pair in "yeast 1" reduced fitness significantly more than deleting their singleton counterpart in "yeast 2" (Qian & Zhang, 2014), that provides further evidence on a genomic level for the role of gene duplication in organismal adaptation.

1.3.2 Mechanisms of replication

To understand the processes which led to the remarkable achievements during primate evolution like the expansion of the neocortex or the occurrence of three-color vision, it is not only necessary to know what but also how this happened.

Therefore, some common mechanisms of gene duplication are introduced in the following.

If homologous sequences of non-sister chromatids do not precisely pair at the end of prophase I during meiosis unequal crossing over can occur. In this case parts of a gene, entire genes or even several genes including their introns can be transferred from one chromatid to the other in a tandem duplication (Zhang J, 2003). Hence, the duplicated genes of those segmental duplications are linked in a chromosome.

Interestingly, an explosion of segmental duplications during primate evolution was observed, very likely at least in part caused by a rapid proliferation of Alu elements about 40 mya, as they are strikingly enriched at the junctions of duplicated genes (Bailey et al., 2003). With more than 1 million copies, these Alu elements constitute about 10% of the human genome (Szmulewicz et al., 1998; Lander et al., 2001).

But also, other kinds of segmental duplications are observed not related to homology-driven crossing over. Replication-dependent chromosome breakages seem to play a role in the generation of tandem duplications because duplication breakpoints are enriched at replication termination sites (Koszul et al., 2004). Also, one should be aware that original tandem arrangements of segmental duplications can be broken up by subsequent rearrangements. Thus it could be assumed that tandem arrangements represent more recent duplication events (Friedman & Hughes, 2003).

Another widespread mechanism of duplication is the retrotransposition of a gene (Torrents et al., 2003). This procedure is performed by a retrovirus that integrates reverse transcribed mature RNAs at a random site in the genome. These duplication sites are conspicuous since the involved genes lack introns and possess a poly-A tail instead. Moreover, they are most often separated from their regulatory elements.

Consequently, these newly integrated sequences only rarely give rise to expressed full-length coding sequences and usually directly become pseudogenes (see: 1.3.3). However, deviations from these patterns can occur, for example when the sequence is inserted downstream of a promoter sequence by chance (Long, 2001). Besides, it is improbable to have blocks of genes duplicated together by retrotransposition, as long as they are not all in one operon (Zhang J, 2003; Hurles, 2004).

A common mechanism of gene duplication in plants and a critical factor in early vertebrate evolution is the whole genome duplication resulting from the nondisjunction of all chromosomes during meiosis (McLysaght et al., 2002; Van de Peer et al., 2003). As one can imagine it results in a massive chance for a step-change in organismal complexity. Albeit, in metazoan, it is a rare event since it comes with a substantial problem for the faithful transmission of a genome from one generation to the next (Hurles, 2004).

If duplications of whole chromosome sets could occur, of course, also aneuploidy as the nondisjunction at a single chromosome can result in an abnormal number of chromosomes and hence, also gene copies. In mammals, aneuploidy often is harmful and usually leads to spontaneous abortions (Jia et al., 2015). But even in the case of viable individuals it still alters the gene dosage in detrimental ways to the organism. Thus it is unlikely to spread throughout populations. Another briefly mentioned mechanism, especially for short sequence duplications, is the replication slippage, which occurs due to errors in DNA replication (Chen et al., 2005).

1.3.3 Fates of duplicated genes

Generally, four potential fates of duplicated genes can be distinguished (Qian & Zhang, 2014): pseudogenization (Ohno, 1970), (Zhang J, 2003), neofunctionalization (Ohno, 1970; Hughes, 1994; He & Zhang, 2005), subfunctionalization (Ohno, 1970; Hughes, 1994; Force et al., 1999) and functional conservation (Zhang J, 2003; Zhang P et al., 2003).

The first obstacle a duplicated gene has to overcome to become visible in evolutionary comparisons is the fixation within the genome. This is a fairly rare event even for new mutations with immediate selective advantage (Kimura, 1979; Lynch & Conery, 2000).

But even when the new gene is fixated it does not has to result in a functional protein. Often mutations destroy a protein's functionality, because of a premature stop codon or the demolition of a major protein domain. Those genes that derived from a functional DNA sequence, but incurred mutations that destroyed their functionality are also called pseudogenes. The likelihood of those mutations increases over time. Often there is only a relatively narrow time window for evolutionary exploration before degradation becomes the most likely outcome (Lynch & Conery, 2000).

Nevertheless, as in the case of the human olfactory receptor gene family, pseudogenization can also occur quite late. In contrast to mice, where only about 20% of the olfactory receptor genes are non-functional, in humans, more than 60% ended up as pseudogenes (Rouquier et al., 2000) – a process that might have been compensated by the better three-color vision in the primate lineage (Zhang P et al., 2003).

As already introduced above, gene duplication creates genetic redundancy. Hence, duplicated genes can accumulate mutations faster than single-copy genes. These mutations do not have to lead to a non-functionalization but can also bear a novel function in a process called neofunctionalization. Those functions can be closely related to the original function, like in the case of the opsin genes (Asenjo et al., 1994), but also very different, like the anti-freeze genes of a widespread Antarctic fish taxon that arose from a duplicated digestive enzyme gene. (Chen et al., 1997; Conant & Wolfe, 2008). Neofunctionalization can occur at the moment of duplication due to a partial duplication or because of the emergence of a chimeric gene when the duplicated gene is inserted into another locus (Lynch & Katju, 2004; Kaessmann, 2010). Of course, it can also be a long process gradually occurring after the entirely correct duplication of a gene (Ohno, 1970; Kimura & Ohta, 1974; He & Zhang, 2005).

Another likely fate of duplicated genes is described as subfunctionalization (Force et al., 1999). In this case, the original function is not merely retained by one copy while the other one either evolves a new feature or degrades but is portioned between the duplicates. This can be an important adaptation for the stable maintenance of these genes within the genome because two genes with similar functions are unlikely to be fixed in a genome unless the extra amount of gene product is advantageous in some way for the fitness of the organism.

Subfunctionalization can occur in several ways. Changes in gene expression are just as conceivable as adaptations on the protein level leading to functional specializations (Zhang J, 2003). While the first case is likely to occur often and soon after duplication (Gu et al., 2002), both can indirectly also contribute to evolutional novelty as it enables duplicated genes to survive in the genome offering a more extended period for an increased chance of neofunctionalization (He & Zhang, 2005).

One of the most popular models to explain subfunctionalization, the *Duplication-Degeneration-Complementation* (DDC) model, was proposed in 1999 (Force et al., 1999). This assumes that different mutations may inactivate different subsets of the original function from each gene duplicate. Hence, though slightly affecting the protein's function, these mutations are neutral for the organisms because the other copy can still compensate for their deleterious effect. Thus, it becomes clear that neofunctionalization and subfunctionalization do not have to be mutually exclusive processes (Fontdevila, 2011).

For some genes, however, there is already an immediate selective advantage of gene duplication through the facilitated elevated expression level: *Duplication for the sake of producing more of the same*, like it has been observed for histones and ribosomal RNA genes (Ohno, 1970). In this case, gene conversion or purifying selection can be advantageous to decrease divergence of the duplicates. Thus, those genes become functionally conserved through the positive effect of increased gene dosage.

In summary, three life stages of a duplicated gene pair can be distinguished. First, it has to be created (creation) resulting in either the loss, which is by far the most common fate or stage two – fixation and preservation with a 100% frequency in a population by natural selection or genetic drift. Finally, a subsequent optimization of the post-fixated gene can occur as introduced above (Conant & Wolfe, 2008) (Fig. 5).

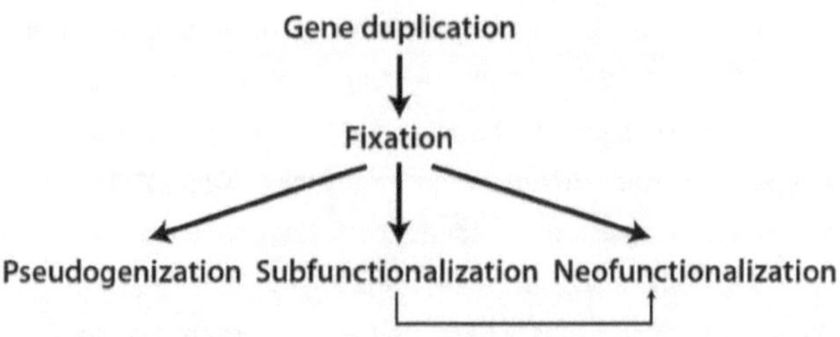

Fig. 5 – Fates of duplicated genes

The first essential step of a duplicated gene to become evolutionary visible is the fixation within the genome. If the protein's functionality is destroyed because of a mutation leading to the demolition of a major functional domain or premature stop codon, the duplicated DNA section becomes a pseudogene. Subfunctionalization describes a process which portions the original function between the duplicates. Neofunctionalization can occur upon duplication due to a mutation or via a subfunctionalized duplicated gene that acquires a novel function.

1.3.4 Which genes tend to duplicate?

There is a high number of transcription factors, kinases as well as particular enzymes and transporters among often duplicated genes. Interestingly, slowly evolving genes tend to be found in duplicates more frequently, but essential genes are not more likely to be duplicated, despite an indirect association between lower rates of evolution and essentiality (Drummond et al., 2006).

So, what could be a reason, that some genes tend to duplicate more often than others? One explanation can be found in the biochemical features of specific proteins which allow them to quickly adapt to new functions, such as the substrate promiscuity of an enzyme, for instance. An opposite approach would be to argue with kind of a duplication resistance of some genes. For example, because of an immediate detrimental effect (Paterson et al., 2006) of the duplication leading to the dosage-balance-hypothesis (Papp et al., 2003). Observations that genes in sparse regions of the protein-protein interaction network and with weaker knockout fitness defects tend to have higher rates of duplications in yeast, because of less potentially detrimental dosage conflicts, support this hypothesis (Conant & Wolfe, 2008).

1.3.5 Human adaptation and gene duplication

Interestingly, there has been a burst of segmental duplications in the lineage of the great apes including humans (Marques-Bonet et al., 2009) with the potential to directly facilitate functional novelty but also the probability for further duplications because of larger and more abundant tracts of identical sequences. Thus, gene duplications are a reasonable cause to consider, when trying to understand drastic changes over short evolutionary periods of times.

Comparative and sequence identity analyses support a threefold excess of duplications in the common ancestor of the human and great ape lineage about 10 mya, in contrast to deletions which occurred rather in the expected frequency (Sudmant et al., 2013; Megan Y Dennis & Eichler, 2016). Intriguingly, the human-specific duplications identified are significantly enriched in genes involved in neurodevelopment (Fortna et al., 2004; Sudmant et al., 2010; Heide et al., 2017).

Thus, uncovering the genetic adaptations and their functional relevance in the lineage of the great apes, and especially humans will offer great insights not only into the understanding of what made us human but also human-specific pathologies.

1.4 Human-specific signatures of neocortical expansion

The human brain might basically conform the primate scaling rules (Herculano-Houzel, 2009), but to explain the human uniqueness, we have to understand the adaptations which occurred after the divergence from the chimpanzee-bonobo lineage 5-7 mya (Glazko & Nei, 2003). Recent reviews from Florio et al. (2017) and Heide et al. (2017) give an excellent overview about those human-specific adaptations based on gene duplication and beyond that were observed to impact neocortical expansion.

1.5 Identification of human-specific genes expressed in the developing neocortex

Based on the premise that the human brain systematically differs from that of other species, it follows naturally that this has to result mainly from a genetic heritage (Vallender, 2012). Hence, an understanding of the changes that occurred in the human genome after the divergence from the chimpanzees is critical for better insights into the molecular pathways underlying human uniqueness in health and disease. Albeit, particularly genes embedded within recently duplicated sequences which might be of exceptional interest in regard to this question (see: 1.3) have been hard to detect in standard genetic analyses (Bailey et al., 2002), since they were frequently misassembled or missing from the reference genome (Bailey et al., 2001;

Heide et al., 2017). The combination of evolutionary developmental biology and next generation sequencing and the advancement of the reference genomes, finally, made it possible to identify the gene-expression counterparts for theoretical concepts like homology (Kalinka et al., 2010; Necsulea et al., 2014) and also enabled rapid progress in the field of neocortical evolution through the novel opportunity of comparative transcriptomics (Florio, 2015; Montiel et al., 2016).

Human neocortical expansion, anon, is mainly driven by an increased proliferative capacity of NPCs in the developing neocortex (Azevedo et al., 2009; Florio & Huttner, 2014; Dehay et al., 2015; Namba & Huttner, 2017; Sousa et al., 2017). Marta Florio and colleagues conclusively used available transcriptome datasets to perform a comprehensive screen for protein-coding genes preferentially expressed in progenitors, but not neurons of the human fetal neocortex. This analysis led to the identification of 15 human-specific genes exhibiting such an expression pattern (Florio et al., 2018). These findings became the foundations of this book since we decided to functionally characterize a gene family constituting 3 of the 15 genes: *FAM72B, FAM72C,* and *FAM72D*. This was based on the exciting fact that humans acquired these three additional paralogues of the ancestral gene in contrast to all distinct species on the earth sequenced to date as well as promising previous findings of other groups studying the function of the ancestral mouse gene *Fam72a*.

1.6   Family with sequence similarity 72 (FAM72)

1.6.1   Evolutionary origin

After the divergence from the chimpanzee lineage, the *Family with sequence similarity 72 (FAM72)* originated from the ancestral *FAM72A (LMPIP, p17, Ugene-p)* locus on chromosome 1. Most likely three duplication events in conjunction with the duplications of the genomic neighbor *SRGAP2* in the human lineage brought up the three human-specific paralogues *FAM72B, FAM72C and FAM72D (Ugene-q)* (Kutzner et al., 2015). Thus, insights about the evolutionary history of *SRGAP2A, B, C & D* might be closely related to *FAM72A, B, C & D*. As already mentioned above (see: 1.3.2) Alu repeats are known to be highly associated with primate genomic duplications (Bailey et al., 2003). Interestingly, Dennis et al. (2012) found in that these repetitive elements might have also been involved in the duplication of the *SRGAP2* and hence, the *FAM72* genes. Furthermore, assuming a divergence from the chimpanzee lineage 6 mya, this study estimated the three duplication time points. The first duplication of *SRGAP2A* (1q32.1) resulted in *SRGAP2B* (1q21.1) and happened about 3.4 mya and was followed by a duplication of the new *SRGAP2B* locus leading to *SRGAP2C* (1p12) about 2.4 mya. Finally, the *SRGAP2B* locus duplicated once more within 1q21.1 and led to *SRGAP2D* 1 mya (Dennis et al., 2012) (Fig. 6). This data and the genomic location of the four *FAM72* paralogues suggest that the gene duplication of *FAM72D* occurred 3.4 mya, *FAM72C* 2.4 mya and *FAM72B* 1 mya (Fig. 6).

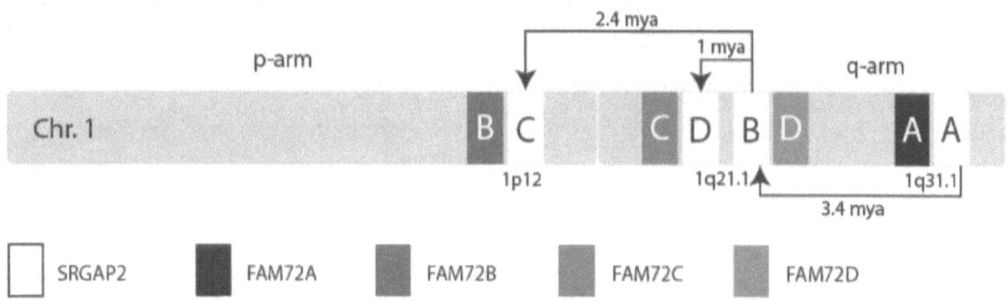

Fig. 6 – Genomic localization and estimated duplication time points

The figure schematically shows the location of *SRGAP2 A, B, C* and *D* as well as of *FAM72 A, B, C* and *D* on chromosome 1 (Chr. 1) within in the human genome. Dennis and colleagues calculated that the duplication events leading to the human specific paralogues of FAM72A and SRGAP2 occurred 3.4 - 1 million years ago (mya). The evolutionary history of the duplication events is traced by the black arrows. The figure is adapted from Dennis et al. (2012) (Fig. 5) and Kutzner et al. (2015) (Fig. 1A).

1.6.2 Subcellular localization

To date, four different studies attempted to determine the subcellular localization of *FAM72A* in various cell lines. Guo et al. (2008) constructed a V5-tagged *FAM72A* (*Ugene-p*) and overexpressed it in SW480 cells (a colon cancer cell line). They found that it accumulates in the nucleus, which was defined by a 4′,6-Diamidino-2-phenylindole (DAPI) staining. As the protein sequence of *FAM72A* lacks a nuclear localization sequence (NLS), they concluded that it might be held in the nucleus by its interaction with other nuclear proteins (Guo et al., 2008). Nehar et al. (2009) performed immunocytochemistry (ICC) in H9C2 (cell line from rat heart myoblast) cells that exhibit a high endogenous FAM72A (p17) expression. In contrast to Guo et al., they describe the observed expression pattern of FAM72A to be different in comparison to DAPI (nucleus) but also vimentin (cytoskeleton). To further dissect this observation a *GFP*-tagged *FAM72A* was transfected to CHO (Chinese hamster ovary) cells showing a localization in either membrane or cytosol, but not the nucleus. Subfractionation of H9C2 cells finally showed that *FAM72A* might be in the cytosol and the membrane (Nehar et al., 2009). Another ICC in H9C2 cells was performed by Heese et al. (2013) showing a cytoplasmic and to a lesser extent nuclear localization. Furthermore, they predicted a transmembrane domain using a biocomputational analysis tool (Heese, 2013) (see: Results I). Wang et al. subfractionated HEK 293 cells (with endogenous FAM72A expression) into four parts and performed Western Blots with a polyclonal anti-FAM72A (LMPIP) antibody on these fractions showing a band in the organelle/membrane fraction, but not cytoplasm, nucleus or cytoskeleton. Immunofluorescence analysis revealed a dot-shaped pattern and a less dot-shaped distribution of aggregates more co-localized with organelles such as mitochondria, but not endoplasmic reticulum and Golgi body (Wang et al., 2011). A subcellular localization associated to mitochondria would be particularly interesting in conjunction with the findings discussed in 7.3.5. Albeit, it remains an open challenge to determine the subcellular localization of the FAM72 proteins reliably. Therefore, it might be necessary to differentiate between cell types or developmental stages.

1.6.3 Cell cycle regulation

FAM72A (LMPIP, p17 or Ugene-p) was identified to be induced upon Epstein-Barr-virus (EBV) infection, which is associated to various human neoplasms such as nasopharyngeal carcinomas (Wang et al., 2011). Cancer cells are often characterized by an altered cell cycle. Therefore, the effect of FAM72A on the cell cycle was studied. After transfection of *GFP-FAM72A* in EBV-negative nasopharyngeal carcinoma cell (NPCC) lines, FACS was performed to select the GFP+ cells. These cell populations overexpressing FAM72A showed a decreased G1-phase compared to the vector control (From 53% vector control to 38% FAM72A) and a little increase in the apoptotic cell population and the multiploid stage (Wang et al., 2011), whereas no relevant change in the S and G2/M phase could be observed, suggesting that FAM72A shortens the G1-phase. This assumption is supported by the findings of enhanced levels of cell cycle activators like cyclin D1, CDK2, CDK4, and E2F1. Furthermore, FAM72A transfected TW01 cells showed an increased proliferation compared to the mock control and vice versa effects in the case of *FAM72A* knock down driven by RNAi (Wang et al., 2011).

1.6.4 NPC maintenance

The trimethylation of histone H3 at lysine 4 (H3K4me3) is associated to the transcription start sites of active genes (Santos-Rosa et al., 2002; Barski et al., 2007; Benayoun et al., 2014; Guo et al., 2014) and preferentially marks genes which are essential for the function and identity of a cell type (Ang et al., 2011; Schmitz et al., 2011). This characteristic of the chromatin modification H3K4me3 was used to test whether the 5% broadest H3K4me3 domains could be used as a tool to uncover novel cell regulators in a particular cell type such as NPCs. Therefore, Benayoun et al. isolated NPCs from the adult mouse subventricular zone. The top 5% broadest H3K4me3 domains were enriched for genes known to be NPC regulators. Interestingly, Fam72a was amongst the genes with the broadest H3K4me3 domains (Benayoun et al., 2014) but not implicated in the maintenance of NPCs, hitherto (Guo et al., 2008; Nehar et al., 2009; Wang et al., 2011; Kutzner et al., 2015). To test the functional relevance of this discovery, candidate genes, including *Fam72a*, were knocked down (KD) using a lentiviral-based RNA interference approach (shRNA) in the primary adult NPCs. The *Fam72a* KD cells showed a decrease in proliferation but an increase in neuronal differentiation. This suggests that the protein may restrain the differentiation of the adult NPCs (Benayoun et al., 2014).

2   Aims & approaches

This study aims to investigate the role of two genes which are expressed in neural progenitor cells (NPCs) of the human fetal neocortex: the human-specific *FAM72D* and the ancestral *FAM72A* regarding the evolutionary expansion of the human neocortex. The focus is to examine the effects of these genes on the identity and proliferative capacity of embryonic mouse NPCs and to analyze differences in the transcriptional signature upon their ectopic expression in the developing mouse neocortex.

Hypotheses

I)

    a. *The forced expression of FAM72D in the developing mouse neocortex leads to an increased proliferative capacity of the targeted NPCs.*

    b. *The amino acid substitutions in FAM72D support an increased proliferative capacity of mouse NPCs compared to FAM72A.*

This is approached by in utero electroporation of plasmid DNA to drive the expression of FAM72A and FAM72D in mouse NPCs in different embryos of the same litter.
After that potential changes in the proliferation of NPCs are investigated as an indicative parameter for cortical expansion during development.

II)

The forced expression of FAM72A or FAM72D in the developing mouse neocortex causes a distinctive gene-expression signature between the ancestral and the human-specific paralogue.

Analogous to the experiments on the protein level (Hypotheses I), control, *FAM72A,* and *FAM72D* DNA plasmids are in utero electroporated and co-expressed with a fluorescent protein. Upon microdissection and cell dissociation, the electroporated cells are isolated by fluorescent-activated cell sorting (FACS). In the following, RNA sequencing and transcriptome analysis are performed on the isolated cells to compare the gene expression differences between the control, ancestral FAM72A and human-specific FAM72D condition.

3 Results I

3.1 From genes to proteins: 1 family – 4 paralogues

In contrast to every other species that genomes have been sequenced, in humans, the *FAM72* gene family consists of four paralogues. In case of the long isoforms, the mature mRNA of all four genes consists of 450 nt and transcribes a 149 amino acids (aa) long protein of 16-17 kDa (*FAM72A* Ensembl ID: ENSG00000196550). This is why FAM72A was previously also called p17 (Nehar et al., 2009). At the cDNA level, the human *FAM72A* is identical to the chimpanzee *FAM72A*. Hence, there is no difference between the FAM72A protein sequences of these two species, and *FAM72A* will be considered as the ancestral paralogue of the gene family in the following (Fig. 7). In total, the human-specific paralogues differ at six positions from *FAM72A*. Albeit, the substitution at nucleotide position 6 in *FAM72C* and *D* is a silent mutation. Consequently, only five out of six nucleotide substitutions, result in changes at the protein level (Fig. 9). Two mutations occurred in the exons of *FAM72B*: the first at nucleotide 281 where a thymine (T) to cytosine (C) mutation caused the switch from leucine to proline (aa 94) in FAM72B, which might lead to a change in the 3D conformation of the protein and the second one at nucleotide 364 changing the nonpolar glycine in FAM72A to the nonpolar valine (aa 122). Two additional mutations at nucleotide 295 and 373 are shared by *FAM72C* and *FAM72D* affecting the biochemical properties of the proteins, since the nonpolar, hydrophobic glycine (aa 99) and tryptophan (aa 125) got exchanged against the hydrophilic, basic arginine in FAM72C and D. Finally, FAM72C is different from FAM72D in aa position 82, where FAM72D is identical to FAM72A but FAM72C has a valine instead of the ancestral glycine (Figs. 7 and 9).

The comparison of the human FAM72A and mouse Fam72a protein sequences reveals that 15 out of 149 aa are different between mouse and human. Four of 5 aa that are mutated in FAM72B, C or D, are conserved between mouse and human Fam72a/FAM72A.

```
A
          1                                                            60
Fam72a    MSTNNCTFKDRCVSILCCKFCKQVLSSRGMKAVLLADTDIDLFSTDIPPTNTVDFIGRCY
FAM72A    MSTNICSFKDRCVSILCCKFCKQVLSSRGMKAVLLADTEIDLFSTDIPPTNAVDFTGRCY

                                82            94    99              120
Fam72a    FTGICKCKLKDIACLKCGNIVGYHVIVPCSSCLLSCNNGHFWMFHSQAVYGINRLDATGV
FAM72A    FTKICKCKLKDIACLKCGNIVGYHVIVPCSSCLLSCNNGHFWMFHSQAVYDINRLDSTGV

          122 125     134           149
Fam72a    NLLLWGNLPETEECTDEETLEISAEEYIR        [ ] conserved   [ ] not conserved
FAM72A    NVLLWGNLPEIEESTDEDVLNISAEECIR


B
          1                                                            60
chimp F72A  MSTNICSFKDRCVSILCCKFCKQVLSSRGMKAVLLADTEIDLFSTDIPPTNAVDFTGRCY
human F72A  MSTNICSFKDRCVSILCCKFCKQVLSSRGMKAVLLADTEIDLFSTDIPPTNAVDFTGRCY
FAM72B      MSTNICSFKDRCVSILCCKFCKQVLSSRGMKAVLLADTEIDLFSTDIPPTNAVDFTGRCY
FAM72C      MSTNICSFKDRCVSILCCKFCKQVLSSRGMKAVLLADTEIDLFSTDIPPTNAVDFTGRCY
FAM72D      MSTNICSFKDRCVSILCCKFCKQVLSSRGMKAVLLADTEIDLFSTDIPPTNAVDFTGRCY

                   74                           95                  120
chimp F72A  FTKICKCKLKDIACLKCGNIVGYHVIVPCSSCLLSCNNGHFWMFHSQAVYDINRLDSTGV
human F72A  FTKICKCKLKDIACLKCGNIVGYHVIVPCSSCLLSCNNGHFWMFHSQAVYDINRLDSTGV
FAM72B      FTKICKCKLKDIACLKCGNIVGYHVIVPCSSCLPSCNNGHFWMFHSQAVYDINRLDSTGV
FAM72C      FTKICKCKLKDIACLKCGNIVVYHVIVPCSSCLLSCNNRHFWMFHSQAVYDINRLDSTGV
FAM72D      FTKICKCKLKDIACLKCGNIVGYHVIVPCSSCLLSCNNRHFWMFHSQAVYDINRLDSTGV

                              149
chimp F72A  NVLLWGNLPEIEESTDEDVLNISAEECIR
human F72A  NVLLWGNLPEIEESTDEDVLNISAEECIR
FAM72B      NILLWGNLPEIEESTDEDVLNISAEECIR
FAM72C      NVLLRGNLPEIEESTDEDVLNISAEECIR
FAM72D      NVLLRGNLPEIEESTDEDVLNISAEECIR
```

Fig. 7 – The protein sequence of FAM72 in mouse, chimpanzee and human

(A) Alignment of the mouse Fam72a and human FAM72A protein sequences; differences in amino acids (aa) between mouse and human are highlighted in the human sequence in red. The blue rectangles indicate aa which are mutated in at least one of the human-specific FAM72 paralogues compared to FAM72A, but not between human FAM72A and mouse Fam72a. The single red rectangle marks one of the aa mutated in FAM72B compared to FAM72A and between mouse Fam72a and human FAM72A. The two cysteine residues that differ between mouse and humans at position 134 and 147 are highlighted in bold.

(B) Alignment of the chimpanzee FAM72A (chimp F72A) and the human FAM72 (F72A) protein sequences. Note that the human and chimpanzee FAM72A sequences are identical. The aa mutated in FAM72B, C or D compared to FAM72A are highlighted in red.

3.2 *FAM72* mRNA expression levels in the developing mouse and human neocortex

The mRNA sequencing datasets of Fietz et al. (2012) and Florio et al. (2015) were used to analyze the gene expression levels of *FAM72A, B, C* and *D* in humans or of *Fam72a in mice* in NPCs and neurons. Fietz and colleagues used a laser dissection microscope to isolate the

VZ, iSVZ, oSVZ and cortical plate (CP) from 13–16 week post-conception (wpc) human fetuses and E14.5 mouse embryos. Afterward, an RNA sequencing on the cells contained in each of the germinal zones (VZ, iSVZ or oSVZ) and the CP was performed (Fietz et al., 2012). Florio and colleagues developed a novel method to isolate the aRG, bRG, bIP based on their morphological properties from the human (wpc 13) and mouse (E14.5) developing neocortex. To isolate neurons either transgenic Tubb3 – green fluorescent protein (GFP) mice embryos or vital DNA staining with a fluorescent dye of the cell suspension (for fetal human neocortex) were used to distinguish neurons based on its fluorescent staining (GFP, mice) or due to the different DNA content in G_0 compared to NPCs which are in S-G_2- or M-phase. Finally, they used fluorescence-activated cell sorting to isolate the different cell populations and to secondly conduct an RNA sequencing to analyze the transcriptome of these cells in mice and humans (Florio et al., 2015). Both datasets show an expression pattern of all FAM72 paralogues almost exclusively restricted to the GZ and NPCs (Fig. 8).

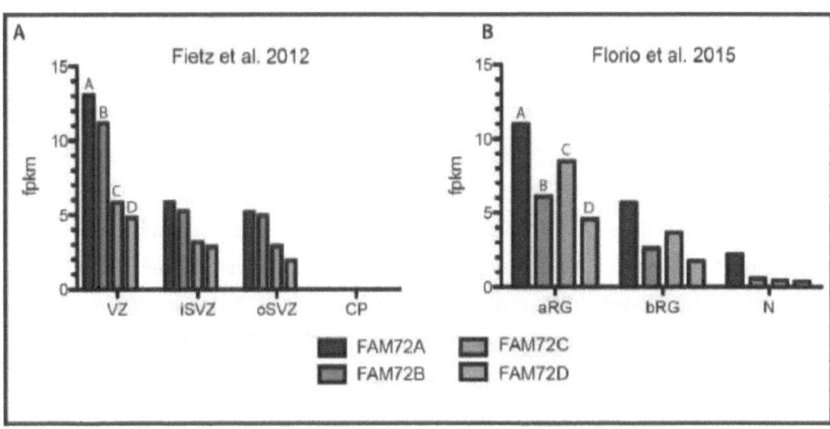

Fig. 8 – FAM72 mRNA expression levels in the developing human neocortex

The mRNA sequencing datasets of Fietz et al. (2012) (A) and Florio et al. (2015) (B) were used to analyze the gene expression levels of *FAM72A, B, C* and *D* (indicated with A, B, C and D above the VZ and aRG bars) in the germinal zones (VZ, iSVZ and oSVZ) and cortical plate (CP) (A) or NPCs and neurons (B). Both datasets show an expression pattern of all FAM72 paralogues almost exclusively restricted to the GZ or NPCs.

3.3   Computational analyses

3.3.1 Proportion of cysteines

As part of the sequence analysis, I realized a strikingly large proportion of cysteines in the amino acid sequence. In fact, almost 10% (14 / 149 aa, 9,4%)[1] of all aa constituting the FAM72 proteins are cysteines which is about 3-4 times more than the average frequency observed in mammalian proteins (Gaur, 2014) and might be at the same time at least in part explained in the relatively small size, since smaller proteins are more dependent on stabilizing disulfide bonds (White, 1992). The total number of cysteines is conserved between mouse and human, although humans have an additional one at aa position 147 and mice at aa position 134 (Fig. 7). Taken together, the relatively higher proportion of cysteines will be of importance to better understand the biochemical properties of the protein, but is not further discussed in this book.

3.3.2 Transmembrane domain

A transmembrane domain (TMD) (aa 74-95, Fig. 7) of FAM72A was predicted using a biocomputational analysis tool (http://harrier.nagahama-i-bio.ac.jp/sosui/ sosui_submit.html) (Heese, 2013). Therefore, we used the same tool to test if it also predicts such a domain for FAM72B, C and D. Surprisingly, FAM72B and FAM72C were calculated to be soluble proteins in contrast to FAM72A and FAM72D that contain a TMD. In addition to this TMD prediction tool, we also used another TMD analysis tool provided by the Center for biological analysis (http://www.cbs.dtu.dk/services/TMHMM/). As a result, neither FAM72A, B, C or D was predicted to have a TMD.
Regarding these controversial TMD predictions and previously published results described above, subcellular localization of Fam72a and its human orthologues is still under discussion. We also attempted to examine the subcellular localization of the human FAM72 proteins by immunofluorescence. However, there were no conclusive signals because of the poor sensitivity and specificity of antibodies (see: 9.8).

---

[1] The cysteine proportion of 9,4% is also confirmed by Kutzner and colleagues who previously published the false cysteine proportion of 8,7% (Kutzner et al., 2015).

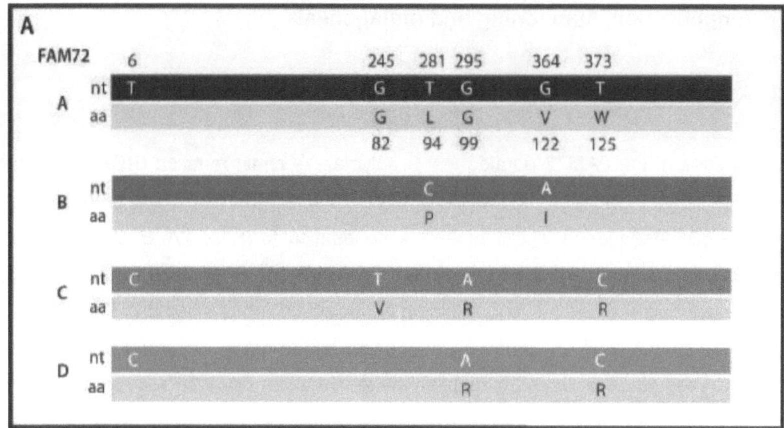

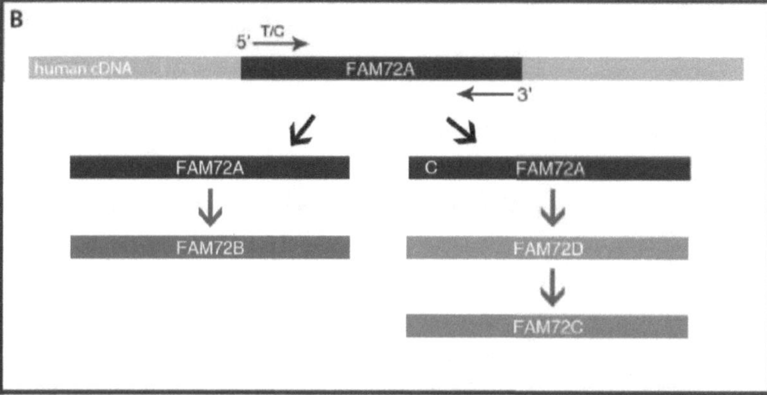

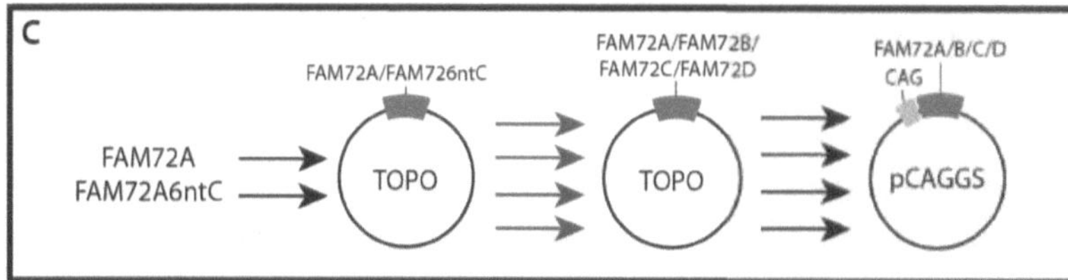

Fig. 9 – Cloning and generation of the FAM72 constructs

(A) Summary of all mutations within the exons of the four FAM72 paralogues; numbers above FAM72A bar indicate nucleotide (nt) position and below the amino acid (aa) position of the specific mutation in FAM72B, C, or D compared to FAM72A; no letter means it is identical to FAM72A at this position.

(B-C) Workflow FAM72A, B, C, D cloning; (B) Amplification of FAM72A from human fetal (GW 12) neocortical cDNA by PCR with two different 5' primers and one 3' primer (blue arrows) led to FAM72A and FAM72A6ntC. The PCR products were used for mutagenesis (green arrows) to produce FAM72B, C and D.

(C) FAM72A and FAM72A6ntC were cloned into TOPO vector before mutagenesis (green arrows), afterwards TOPO-FAM72A/B/C/D was subcloned into pCAGGS vector that contains a CAG promotor (light green) to obtain pCAGGS-FAM72A/B/C/D vectors.

3.4   Amplification, subcloning and mutagenesis

3.4.1 Amplification from human cDNA

To obtain DNA of the FAM72 paralogues, a polymerase chain reaction (PCR) on a human brain cDNA sample from gestation week 12 (GW 12) was performed using primers targeting the *FAM72* cDNA sequence. The 3' primers were identical for *FAM72A, B, C,* and *D*, but two different 5' primers had to be used to amplify *FAM72A/B* and *FAM72C/D* to produce a clean sequence including the silent T to C mutation in nucleotide (nt) position 6 in *FAM72C* and *D*. To verify the propriety of the amplified sequences all amplification products were Sanger sequenced. We found that the PCR resulted either in a clean *FAM72A* sequence when the 5' primer for *FAM72A/B* was used and a *FAM72A* sequence with the nt 6 T to C mutation (*FAM72A6ntC*) induced by the *FAM72C/D* 5' primer. None of the samples sequenced comprised *FAM72B, C* or *D* (see: 7.3.3)

3.4.2 Verification of the pCAGGs vectors [2]

Before examining the role of *FAM72* paralogues in the NPCs, several preliminary experiments were conducted. In the first place, the coding sequences of the expression plasmids (pCAGGs-*FAM72A/B/C/D*) were confirmed to be matched with the reference sequences obtained from Ensembl (see: 9.2.1) by Sanger sequencing.

Second, the protein expression of the plasmids was confirmed by immunoblotting of cell lysates obtained from Cos7 cells (African green monkey kidney cells) that were transfected either with pCAGGs-*FAM72A/B/C/D* or an empty pCAGGs vector as a control. Afterward, an immunoblot (IB) with a primary antibody recognizing all four *FAM72* paralogues was performed to test the presence of the FAM72 proteins in the transfected Cos7 cells. The result clearly shows bands at the expected size of the FAM72 proteins between 16-17 kDa for each *FAM72* construct, but none in the lane of the Cos7 cells only transfected with an empty pCAGGs vector (Fig. 10).

Lastly, to further test if the construct also expresses FAM72 in the embryonic mouse brain, plasmids encoding FAM72A and FAM72D were in utero electroporated into mouse neocortex at E13.5. The brains were dissected 48 h later at E15.5, fixated and cryosectioned afterward. Then we examined the presence of the FAM72 mRNA in the electroporated cortical area via in situ hybridization (ISH) with DNAse treatment to exclude a signal caused by the pCAGGs-

---

[2] In collaboration with Michael Heide (ISH) and Takashi Namba (IB), both MPI CGB, Huttner Lab

FAM72 plasmid. Indeed, in situ hybridization showed a strong signal of the *FAM72* mRNA restricted to the electroporated area (Fig. 10).

Therefore, I concluded that the plasmids lead strong expression of FAM72A/B/C/D at mRNA and protein level, in both cultured cells and embryonic mouse neocortex.

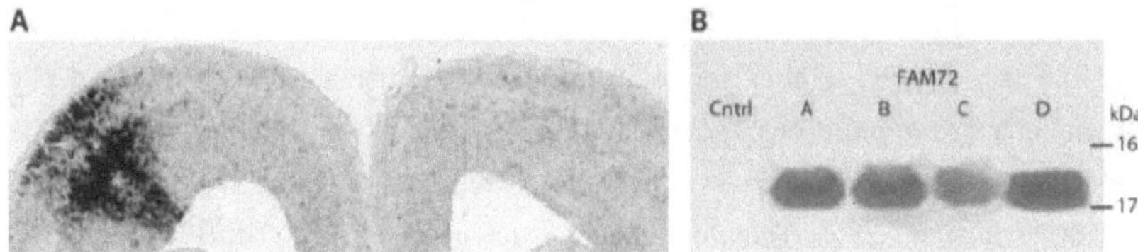

Fig. 10 – Expression of the pCAGGS-FAM72 plasmids

(A) In situ hybridization (ISH) of a mouse neocortex that was electroporated with pCAGGS-FAM72D to prove expression of pCAGGS-FAM72D on the mRNA level. The ISH was conducted in collaboration with Michael Heide. (B) Immunoblot (IB) with primary antibody against FAM72A, B, C or D of Cos7 cells transfected with pCAGGS-FAM72A/B/C/D vectors or pCAGGS empty vectors (Cntrl) shows protein bands of 16-17 kDa as expected for the human FAM72 paralogues. The IB was conducted in collaboration with Takashi Namba.

4.1 Ectopic expression of FAM72A and FAM72D in the mouse dorsolateral neocortex at mid-neurogenesis

Previous studies showed that ectopically expressed human-specific genes induced phenotypical changes in embryonic mouse NPCs in vivo and as a result, the mouse NPCs acquired "humanized" features such as a highly proliferative ability (Florio et al., 2015; Florio et al., 2016; Florio et al., 2018). Therefore, mouse embryonic NPCs can be used as a model system to examine the function of human-specific genes on NPC proliferation and differentiation.

The effect of the *FAM72* gene family could be either just due to an increased gene dosage offered by the three extra paralogues or because of a neofunctionalization of *FAM72B, C* or *D*. To examine these two hypotheses two experiments were conducted. One is the overexpression of FAM72A to examine its dosage effect. The other is the ectopic expression of FAM72D to investigate its potential neofunctionalization because the two arginine substitutions in FAM72D could change protein-protein interactions. This hypothesis has already been supported in a study showing that FAM72D (Ugene-q) in contrast to FAM72A (Ugene-p) does not bind to uracil DNA-glycosylase 2 (UNG2) (Guo et al., 2008).

The expression of pCAGGS-*FAM72A* and pCAGGS-*FAM72D* was induced in apical progenitors of the developing mouse dorsolateral neocortex via in utero electroporation (IUE) (Fig. 11). As a control, I electroporated littermate embryos with the empty pCAGGS vector already used for the immunoblot for plasmid verification (Fig. 10). Besides, each plasmid was co-electroporated with a reporter plasmid (pCAGGS-*RFP*) expressing a nuclear-targeted red fluorescent protein (RFP) to follow the fate of the targeted APs. We decided to use a co-electroporation instead of an *HA-* or *RFP/GFP*-tagged *FAM72A/D*, to avoid any interference caused by the tag with the biological function of the protein and due to our experiences with the *HA-FAM72A* construct, that could not be detected in the immunoblot performed with the available FAM72 antibodies (data not shown).

For the first set of experiments, IUE of FAM72A and FAM72D was performed at E13.5 (mid-neurogenesis). The brains were dissected 24 h (E14.5, at least one cell cycle) or 48 h (E15.5) later, fixated in 4% PFA and cryosectioned for immunohistochemical analyses (Fig. 29, see: 9.1). There were no apparent changes in the distribution of RFP+ cells upon FAM72A, and FAM72D forced expression at 24 h and 48 h post-IUE.

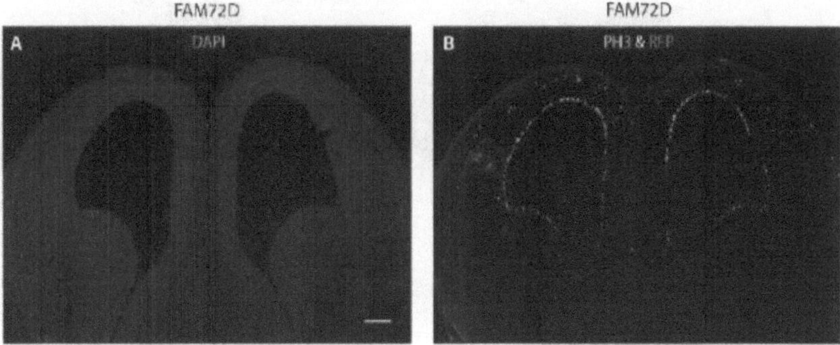

Fig. 11 – In utero electroporation of the dorsolateral mouse neocortex at mid-neurogenesis

Coronal section of an E14.5 mouse brain electroporated at E13.5 (24 h post-IUE).

(A) DAPI staining (blue, cell nuclei).

(B) Double immunofluorescence for PH3 (green, mitotic cells) and RFP (red); RFP indicates the electroporated area (dorsolateral neocortex). Scale bar (150 µm) in A also applies to B.

4.2   NPC proliferation

4.2.1   Assessment of NPC proliferation using Ki67 immunofluorescence

The proliferation was assessed using Ki67, a marker for cycling cells. During interphase, the Ki67 protein is exclusively located in the cell nucleus, while in mitosis it is found at the surface of the chromosomes (Cuylen et al., 2016). Most importantly, it is only present during G_1-, S-, G_2- and M- phase, but not in resting cells (G_0-phase) such as post-mitotic neurons.

Quantification of the total number of Ki67+RFP+ cells per RFP+ cells (from VZ to CP) 48 h upon IUE at E15.5 did not show a significant difference between control (Cntrl) and FAM72D (Fig. 12). To not overlook a slight difference in only one of the germinal zones, I quantified the Ki67+RFP+ cells in the VZ and SVZ-IZ separately. Although a small increase in the SVZ-IZ of brains expressing FAM72D compared to the control was observed, it was not statistically significant (Fig. 12).

We further dissected this observation analyzing the Ki67+RFP+ cells in SVZ-IZ per RFP+ cells in SVZ-IZ 24 h post-IUE including FAM72A and again (Fig. 13), no significant difference between control, FAM72A or FAM72D was detected.

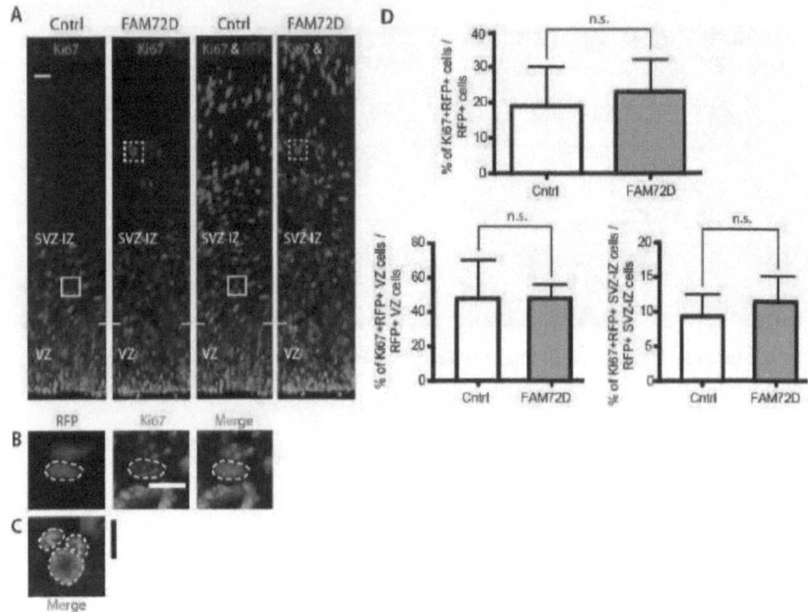

Fig. 12 – Assessment of NPC proliferation using Ki67 immunofluorescence 48 h post-IUE

Mouse neocortex was electroporated at E13.5 with the nuclear-targeting RFP-expressing plasmid either plus control (Cntrl) or FAM72D-expressing plasmids and analyzed at E15.5.

(A-C) Double immunofluorescence for Ki67 (green) and RFP (red). Solid and dashed boxes indicate areas in the SVZ-IZ that are shown at higher magnification in B (nucleus surrounded by dashed lines is Ki67+RFP+ double positive) and C (nucleus surrounded by dashed lines is only Ki67 positive). White lines indicate borders between VZ and SVZ-IZ. Scale bar: 20 µm (A), 10 µm (B, C).

(D) Quantification of the percentage of total RFP+ cells that are Ki67+ (top) and the percentage of RFP+ cells that are Ki67+ in VZ (bottom left) and SVZ-IZ (bottom right). There is no significant difference between control (Cntrl) and FAM72D. Error bars indicate SD; n.s. = not significant. Data are the mean of four independent experiments.

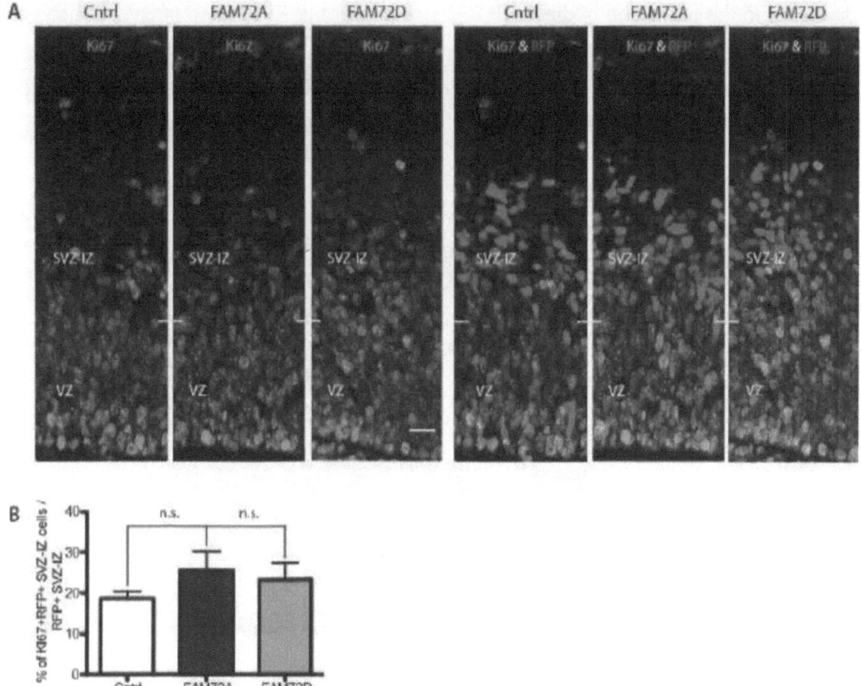

Fig. 13 – Assessment NPC proliferation using Ki67 immunofluorescence 24 h post-IUE

Mouse neocortex was electroporated at E13.5 with the nuclear-targeting RFP-expressing plasmid either plus control (Cntrl), FAM72A- or FAM72D-expressing plasmids and analyzed at E14.5 (24 h post-IUE).

(A) Double immunofluorescence for Ki67 (green) and RFP (red). White lines indicate borders between VZ and SVZ-IZ. Scale bar: 20 µm.

(B) Quantification of the percentage of RFP+ cells that are Ki67+ in SVZ-IZ. There is no significant difference between control (Cntrl), FAM72A and FAM72D. Error bars indicate SD; n.s. = not significant. Data are the mean of three independent experiments.

4.2.2 Cell cycle reentry

Next, I carried out a cell-cycle reentry assay to examine whether the daughter cells enter the cell cycle again. To this end, 5-Ethynyl-2'-deoxyuridine (EdU) was injected at E14.5, 24 h post-IUE (IUE at E13.5) and the brains were harvested and fixated at E15.5 (Fig. 14) followed by Ki67 immunofluorescence. EdU incorporates into the DNA as a thymidine analog during S-phase. Given that the cell cycle length of mouse embryonic NPCs at this stage is shorter than 24 h, the EdU-incorporated cells might exit or reenter the cell cycle at the time point of fixation. Thus, the EdU+Ki67+ cells are the cells that reentered the cell cycle. There was no significant difference in the proportion of the EdU+RFP+ cells over the RFP+ cells in SVZ-IZ between control and FAM72D ectopic expression (Fig. 14).

Quantification of EdU+Ki67+RFP+ cells in the SVZ-IZ per RFP+ or EdU+ cells in SVZ-IZ did not exhibit a significant difference between control and FAM72D sample (Fig. 15).

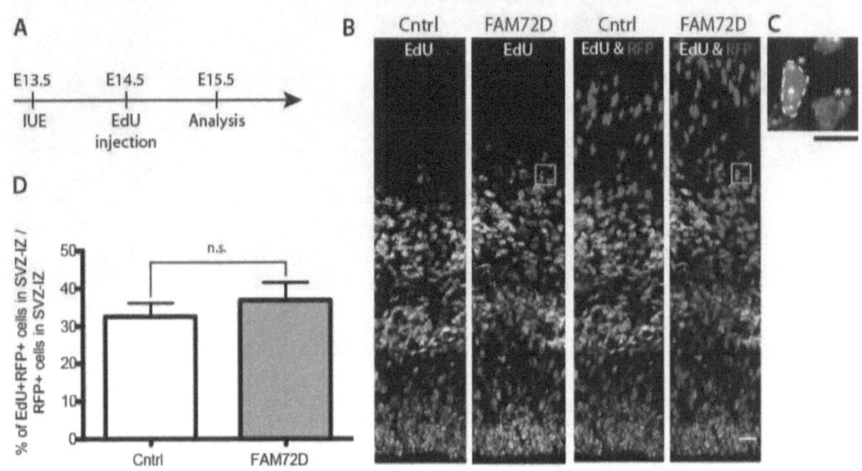

Fig. 14 – Detection EdU incorporation upon expression of FAM72D

(A) Experimental workflow: IUE of the dorsolateral mouse neocortex at E13.5, EdU injection into the abdomen of the electroporated pregnant mouse at E14.5, harvesting of the embryos and further processing for quantitative analysis at E15.5. (B-C) Mouse neocortex was electroporated at E13.5 with the nuclear-targeting RFP-expressing plasmid either plus control (Cntrl) or FAM72D- expressing plasmids. EdU detection (grey) and RFP immunofluorescence 48 h post-IUE was conducted at E15.5. Solid boxes indicate the area in the SVZ-IZ that is shown at higher magnification in C (nucleus surrounded by dashed lines and tagged with * is Edu+RFP+ double positive; nucleus tagged with ** is only EdU positive). Scale bar: 20 µm (B), 10 µm (C).

(D) Quantification of the percentage of RFP+ cells that are Ki67+ in SVZ-IZ. There is no significant difference between control (Cntrl) and FAM72D. Error bars indicate SD; n.s. = not significant. Data shown are the mean of three independent experiments.

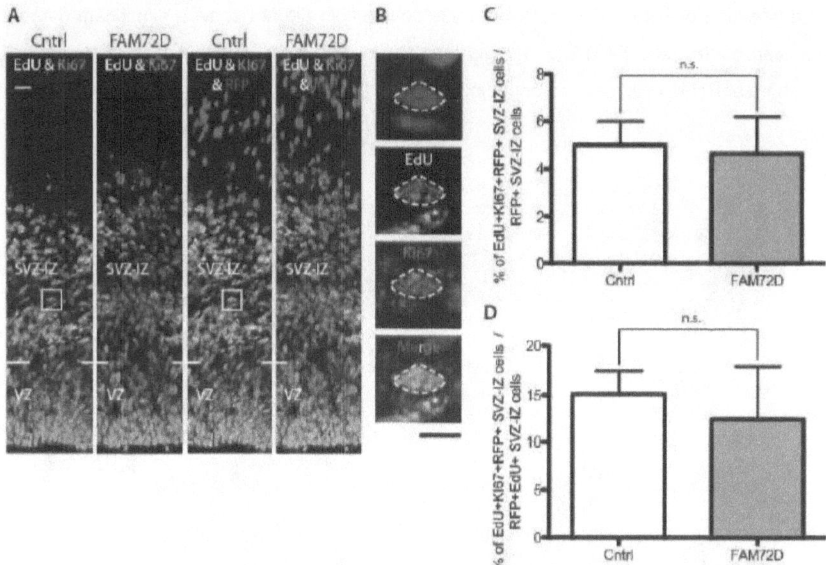

Fig. 15 – Cell cycle reentry

Mouse neocortex was electroporated at E13.5 with the nuclear-targeting RFP-expressing plasmid either plus control (Cntrl) or FAM72D-expressing plasmids and analyzed at E15.5.

(A-B) Double immunofluorescence for Ki67 (green) and RFP (red) plus EdU (grey) detection. The experimental workflow is shown in Fig. 14 A. Solid boxes indicate area in the SVZ-IZ that is shown at higher magnification in B (nucleus surrounded by dashed lines is Ki67+Edu+RFP+ triple positive). White lines indicate borders between VZ and SVZ-IZ. Scale bar: 20 µm (A), 10 µm (B).

(C-D) Quantification of the percentage of RFP+ cells that are EdU+KI67+RFP+ cells in SVZ-IZ and of the percentage of RFP+Edu+ cells that are EdU+KI67+RFP+ cells in SVZ-IZ. There is no significant difference between control (Cntrl) and FAM72D. Error bars indicate SD; n.s. = not significant. Data are the mean of three independent experiments.

4.2.3   Assessment of mitosis using PH3 immunofluorescence

Another way to quantify the number of dividing cells is to observe mitotic cells. Histone H3 is phosphorylated at S10 during mitosis (Hans & Dimitrov, 2001). Thus antibody against phosphorylated histone H3 (PH3) specifically label mitotic cells. To analyze PH3+ cells after FAM72A or FAM72D electroporation, I quantified PH3+RFP+ cells per microscopic field (250 x 300µm).

Quantification of PH3+ cells 24 h post-IUE did not show any significant difference between control, FAM72A or FAM72D. There is a tendency of an increase mainly in the SVZ-IZ of FAM72A expressing brains, though this reflects more an outlier than a general trend (Fig. 16). In addition, I also looked at E15.5, 48 h post-IUE of control and FAM72D-expressing plasmids.

Quantification of PH3+RFP+ cells per microscopic field did not show a significant difference between control and FAM72D. This is also the case for the quantification of the absolute numbers of PH3+ cells per microscopic field (Fig. 17).

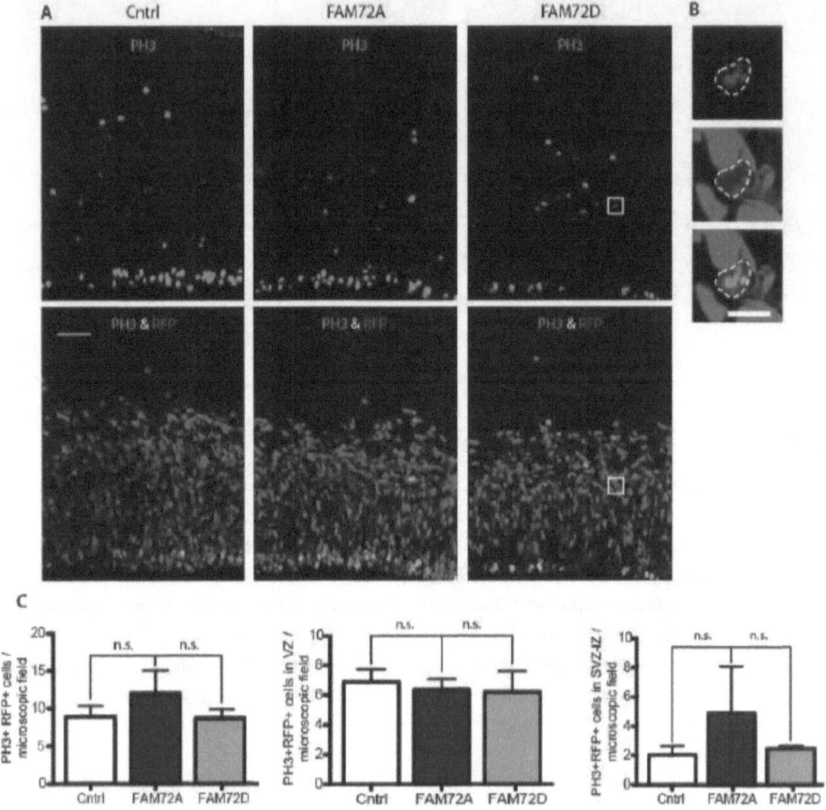

Fig. 16 – Detection of mitoses using PH3 immunofluorescence 24 h post-IUE

Mouse neocortex was electroporated at E13.5 with the nuclear-targeting RFP-expressing plasmid either plus control (Cntrl), FAM72A- or FAM72D-expressing plasmids and analyzed at E14.5.

(A-B) Double immunofluorescence for phosphorylated Histone 3 (PH3) (green) and RFP (red). Solid boxes indicate area in the SVZ-IZ that are shown at higher magnification in B (nucleus surrounded by dashed lines is PH3+RFP+ double positive). Scale bar: 50 µm (A), 10 µm (B).

(C) Quantification of the total number of PH3+RFP+ cells per microscopic field (left) and the number of PH3+RFP+ cells in the VZ (middle) and SVZ-IZ (right) per microscopic field (250 x 300 µm). There is no significant difference between control (Cntrl), FAM72A and FAM72D. Error bars indicate SD; n.s. = not significant. Data are the mean of three independent experiments.

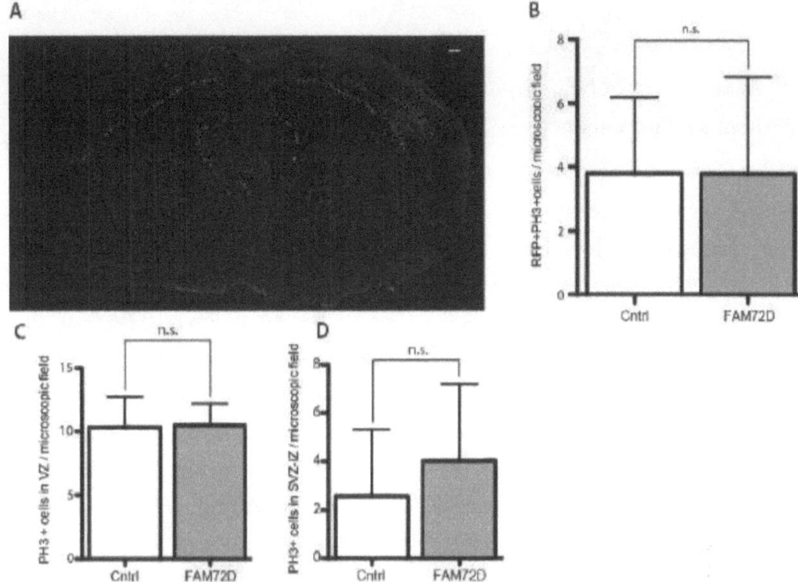

Fig. 17 – Detection of mitoses using PH3 immunofluorescence 48 h post-IUE

Mouse neocortex was electroporated at E13.5 with the nuclear-targeting RFP-expressing plasmid either plus control (Cntrl) or FAM72D-expressing plasmids and analyzed at E15.5.

(A) Double immunofluorescence for phosphorylated Histone 3 (PH3) (green) and RFP (red) of a coronal section of an E15.5 embryonic mouse brain. The right dorsolateral neocortex was in utero electroporated with plasmids expressing FAM72D or nuclear-targeting-RFP. Scale bar: 100 µm.

(B-D) Quantification of the total number of PH3+RFP+ cells per microscopic field (B) and the number of PH3+ cells in the VZ (C) and SVZ-IZ (D) per microscopic field (250 x 300 µm). There is no significant difference between control (Cntrl) and FAM72D. Error bars indicate SD; n.s. = not significant. Data are the mean of four independent experiments.

4.2.4 Conclusion

Taken together, all analyses with a focus on NPC proliferation suggest that neither in utero electroporation of FAM72A nor FAM72D was able to increase the proliferative capacity and cell cycle reentry of NPCs upon their expression at mid-neurogenesis in the developing mouse dorsolateral neocortex.

4.3 NPC abundance

4.3.1 Assessment of NPC abundance using Tbr2 and Sox2 immunofluorescence

Next, it was analyzed whether FAM72A or FAM72D affect the abundance of certain NPC types since a previous study suggested a role of Fam72a in the maintenance of NPC derived from adult mouse SVZ or hippocampus (Benayoun et al., 2014). To this end, I analyzed the number of T-box brain protein (Tbr2, Eomes)+RFP+ (Figs. 18 and 19) or Sox2+RFP+ (Figs. 20 and 21) cells in the FAM72A and FAM72D electroporated developing mouse neocortex. Tbr2 is a marker for BPs (Englund et al., 2005; Kowalczyk et al., 2009) and Sox2 is a marker for RGs (Suh et al., 2007), (Kriegstein & Alvarez-Buylla, 2009; Hansen et al., 2010; Pollen et al., 2015). To assess these parameters, E13.5 mouse embryos were electroporated in utero with the nuclear-targeting RFP-expressing plasmid either plus control plasmid, FAM72A- or FAM72D-expressing plasmids and harvested at E14.5 or E15.5.

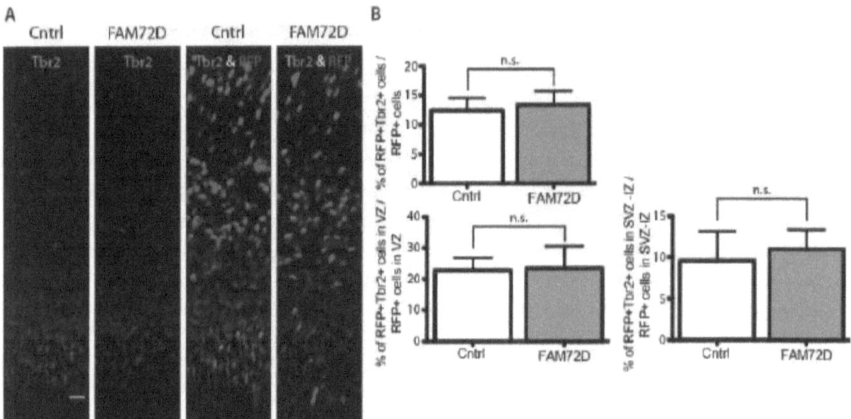

Fig. 18 – Assessment of NPC abundance using Tbr2 immunofluorescence 48 h post-IUE

Mouse neocortex was electroporated at E13.5 with the nuclear-targeting RFP-expressing plasmid either plus control plasmid (Cntrl) or FAM72D-expressing plasmids and analyzed at E15.5.

(A) Double immunofluorescence for Tbr2 (cyan) and RFP (red). Scale bar: 20 µm.

(B) Quantification of the percentage of total RFP+ cells that are Tbr2+ (top) and the percentage of RFP+ cells that are Tbr2+ in VZ (bottom left) and SVZ-IZ (bottom right). There is no significant difference between control (Cntrl) and FAM72D. Error bars indicate SD; n.s. = not significant. Data are the mean of five independent experiments.

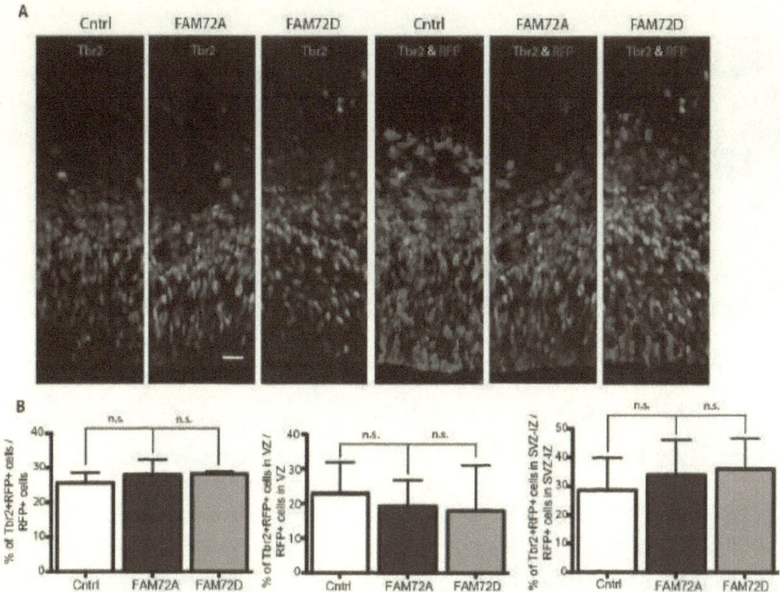

Fig. 19 – Assessment of NPC abundance using Tbr2 immunofluorescence 24 h post-IUE

Mouse neocortex was electroporated at E13.5 with the nuclear-targeting RFP-expressing plasmid either plus control (Cntrl), FAM72A- or FAM72D-expressing plasmids and analyzed at E14.5.

(A) Double immunofluorescence for Tbr2 (cyan) and RFP (red). Scale bar: 20µm.

(B) Quantification of the percentage of total RFP+ cells that are Tbr2+ (left) and the percentage of RFP+ cells that are Tbr2+ in VZ (middle) and SVZ-IZ (right). There is no significant difference between control (Cntrl), FAM72A and FAM72D. Error bars indicate SD; n.s. = not significant. Data are the mean of three independent experiments.

Consistent with the results from the proliferative capacity assay described above compared to control, the proportion of electroporated RFP+ Tbr2+ cells per RFP+ cells showed a significant change upon FAM72 paralogue forced expression in neither VZ nor SVZ-IZ at 24 h (FAM72A and FAM72D) and 48 h (FAM72D) post-IUE (Figs. 18 and 19).

Similarly, the percentage of Sox2+RFP+ cells per RFP+ cells in the VZ and SVZ-IZ did not show significant changes upon FAM72 paralogue forced expression at 24 h (FAM72A and FAM72D) and 48 h (FAM72D) post-IUE (Figs. 20 and 21).

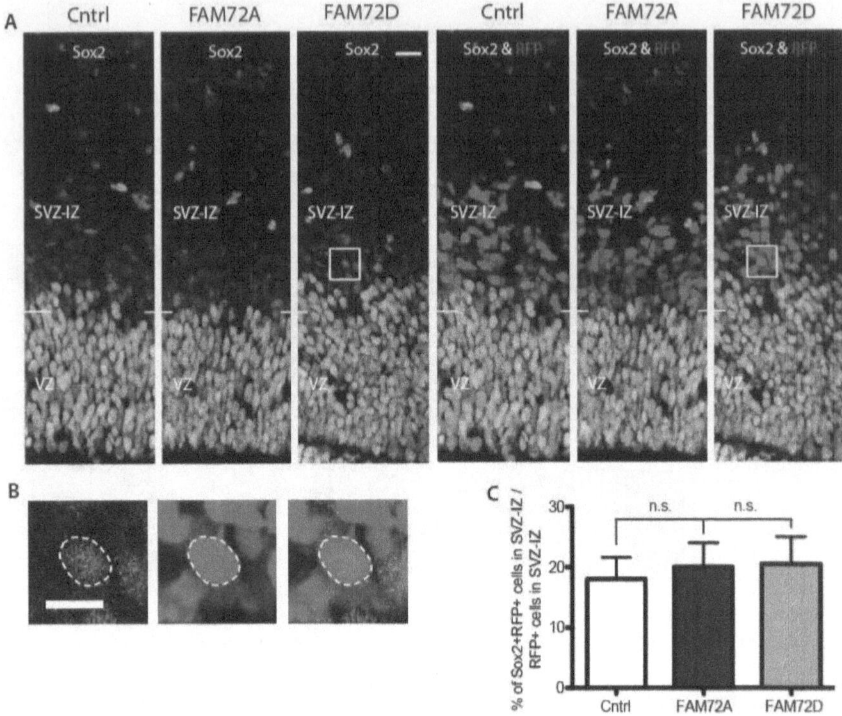

Fig. 20 – Assessment of NPC abundance using Sox2 immunofluorescence 24 h post-IUE

Mouse neocortex was electroporated at E13.5 with the nuclear-targeting RFP-expressing plasmid either plus control (Cntrl), FAM72A- or FAM72D-expressing plasmids and analyzed at E14.5.

(A-B) Double immunofluorescence for Sox2 (yellow) and RFP (red). Solid boxes indicate area in the SVZ-IZ that are shown at higher magnification in B (nucleus surrounded by dashed lines is Sox2+RFP+ double positive). White lines indicate borders between VZ and SVZ-IZ. Scale bar: 20 µm (A), 10 µm (B).

(C) Quantification of the percentage of RFP+ cells that are Ki67+ in SVZ-IZ. There is no significant difference between control (Cntrl), FAM72A and FAM72D. Error bars indicate SD; n.s. = not significant. Data are the mean of three independent experiments.

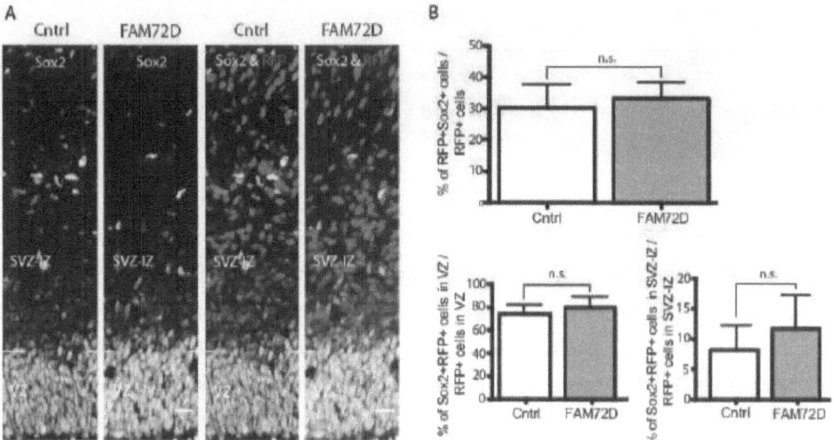

Fig. 21 – Assessment of NPC abundance using Sox2 immunofluorescence 48 h post-IUE

Mouse neocortex was electroporated at E13.5 with the nuclear-targeting RFP-expressing plasmid either plus control plasmid (Cntrl) or FAM72D-expressing plasmids and analyzed at E15.5.

(A) Double immunofluorescence for Sox2 (yellow) and RFP (red). White lines indicate borders between VZ and SVZ-IZ. Scale bar: 20 µm.

(B) Quantification of the percentage of total RFP+ cells that are Sox2+ (top) and the percentage of RFP+ cells that are Sox2+ in VZ (bottom left) and SVZ-IZ (bottom right). There is no significant difference between control (Cntrl) and FAM72D. Error bars indicate SD; n.s. = not significant. Data are the mean of eight independent experiments.

4.3.2 Conclusion

These results suggest that human FAM72A and FAM72D do not significantly increase the proliferative capacity or the pool size of NPC upon forced ectopic expression in the developing mouse dorsolateral neocortex at mid-neurogenesis.

5   Results III

5.1   Ectopic expression of FAM72A and FAM72D in the mouse medial cortex at late-neurogenesis [3]

As introduced in 1.2.2 bRGs of the oSVZ were found to be necessary for the evolutionary expansion of the neocortex in gyrencephalic primates. In mice, these bRGs only exist in small numbers and exhibit significant molecular differences from bRGs in humans, when only the dorsolateral neocortex is regarded. But a recent paper from our lab identified abundant bRGs constituting an oSVZ like zone in the developing mouse medial neocortex at E18.5. (Vaid et al., 2018) The study has further shown that these mouse medial bRGs express many *human bRG-enriched genes,* that are not/or only weakly expressed in mouse lateral bRGs. These findings suggest that the mouse medial bRGs possess features close to human bRGs. Thus, mouse medial bRGs are suitable cells for analyzing the effects of human-specific genes in utero. That is why, control, FAM72A- and FAM72D- expressing plasmids were in utero electroporated at E15.5 to APs of the developing mouse medial neocortex (Fig. 22).

To be able to compare this set of experiments to the results in the dorsolateral neocortex, analogous analyses were performed to assess the abundance of NPC subtypes (Sox2+ and Tbr2+) and NPC proliferation (Ki67). Additionally, gliogenesis was investigated using Olig2 immunofluorescence.

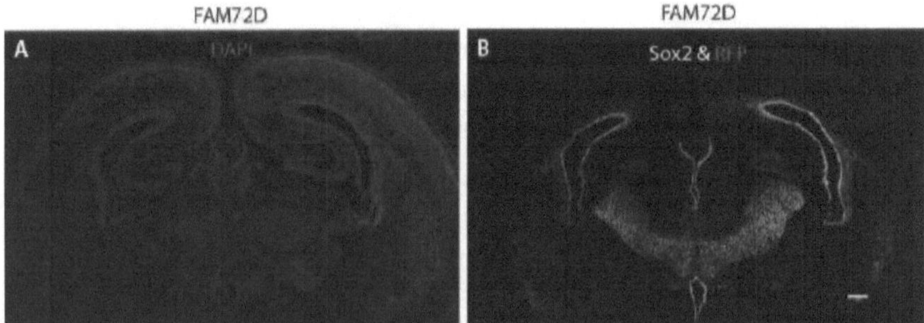

Fig. 22 – In utero electroporation of the medial mouse neocortex at late neurogenesis

Coronal section of an E18.5 mouse brain electroporated at E15.5. IUEs shown in Fig. 22 – 26 were conducted in collaboration with Samir Vaid (MPI CBG).

(A) DAPI staining (blue, cell nuclei).

(B) Double immunofluorescence for Sox2 (yellow, NPCs) and RFP (red). RFP indicates the electroporated area (medial neocortex). Scale bar (200 µm) in B also applies to A.

---

[3] In collaboration with Samir Vaid, Huttner Group MPI CBG

5.2   NPC proliferation - Assessment of the NPC proliferation using Ki67 immunofluorescence

The proliferation in the developing mouse medial neocortex was analyzed using Ki67 immunofluorescence. The percentage of RFP+ cells that are Ki67+ of total RFP+ cells in the entire neocortex was not significantly changed upon ectopic FAM72A or FAM72D expression (Fig. 23).

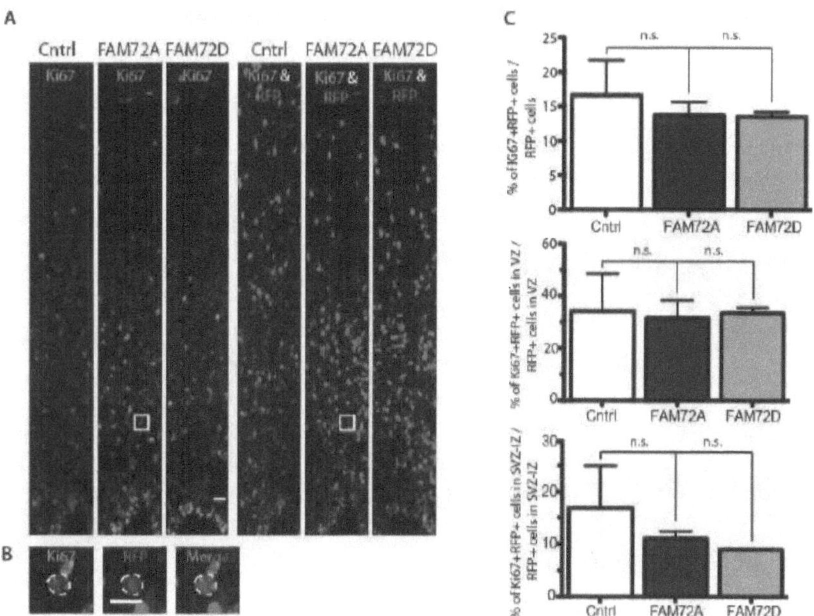

Fig. 23 – Assessment of NPC proliferation using Ki67 immunofluorescence at late neurogenesis

Mouse neocortex was electroporated at E15.5 with the nuclear-targeting RFP-expressing plasmid either plus control (Cntrl), FAM72A- or FAM72D-expressing plasmids and analyzed at E18.5 (72 h post-IUE).
(A-B) Double immunofluorescence for Ki67 (green) and RFP (red). Solid boxes indicate area in the SVZ-IZ that are shown at higher magnification in B (nucleus surrounded by dashed lines is Ki67+RFP+ double positive). Scale bar: 20 µm (A), 10 µm (B).
(C) Quantification of the percentage of total RFP+ cells that are Ki67+ (top) and the percentage of RFP+ cells that are Ki67+ in VZ (middle) and SVZ-IZ (bottom). There is no significant difference between control (Cntrl), FAM72A and FAM72D. Error bars indicate SD; n.s. = not significant. Data are the mean of three independent experiments.

5.3 NPC abundance – Assessment of NPC abundance using Tbr2 and Sox2 immunofluorescence

The abundance of RGs and neurogenic BPs in the medial neocortex at E18.5 upon IUE at E15.5 was evaluated using Sox2 and Tbr2. In the case of Sox2, I quantified the percentage of total RFP+ cells that are Sox2+, as well as the percentage of RFP+ cells that are Sox2+ of RFP+ cells, in VZ or SVZ-IZ. None of these quantifications showed a significant difference between the control, FAM72A and FAM72D brains (Fig. 24). The percentage of total RFP+ cells or RFP+ cells in VZ and SVZ-IZ that are Tbr2+ were not significantly changed among the three groups (Fig. 25).

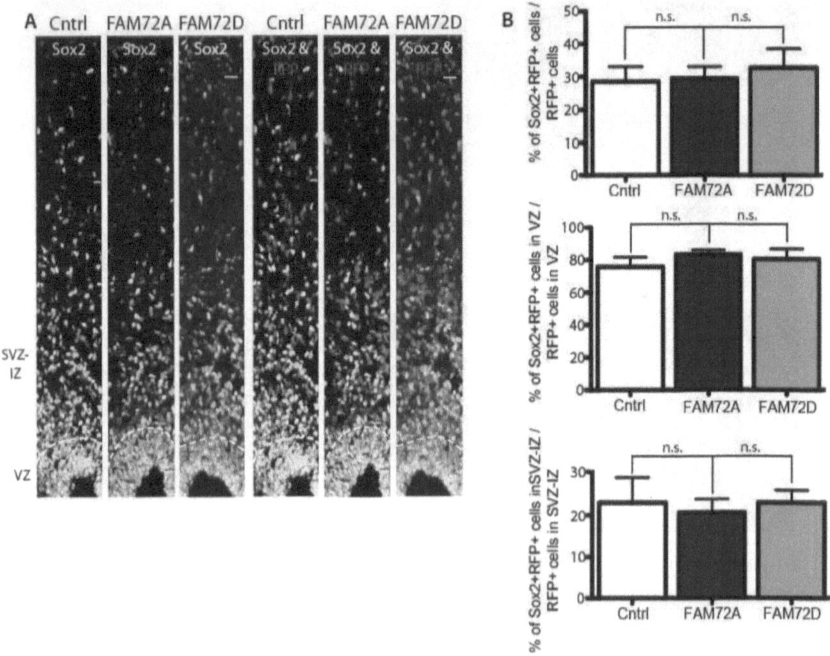

Fig. 24 – Assessment of NPC abundance using Sox2 immunofluorescence at late neurogenesis

Mouse neocortex was electroporated at E15.5 with the nuclear-targeting RFP-expressing plasmid either plus control (Cntrl), FAM72A- or FAM72D-expressing plasmids and analyzed at E18.5 (72 h post-IUE).

(A) Double immunofluorescence for Sox2 (yellow) and RFP (red). Dashed white lines indicate borders between VZ and SVZ-IZ. Scale bar: 20µm.

(B) Quantification of the percentage of total RFP+ cells that are Sox2+ (top) and the percentage of RFP+ cells that are Sox2+ in VZ (middle) and SVZ-IZ (bottom). There is no significant difference between control (Cntrl), FAM72A and FAM72D. Error bars indicate SD; n.s. = not significant. Data are the mean of four independent experiments.

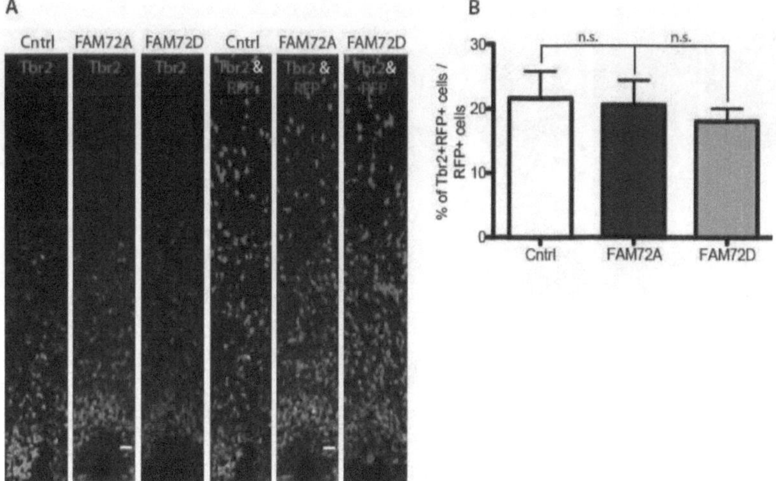

Fig. 25 – Assessment of NPC abundance using Tbr2 immunofluorescence at late neurogenesis

Mouse neocortex was electroporated at E15.5 with the nuclear-targeting RFP-expressing plasmid either plus control (Cntrl), FAM72A- or FAM72D-expressing plasmids and analyzed at E18.5 (72 h post-IUE).

(A) Double immunofluorescence for Tbr2 (green) and RFP (red). Scale bar: 20µm.

(B) Quantification of the percentage of total RFP+ cells that are Tbr2+. There is no significant difference between control (Cntrl), FAM72A and FAM72D. Error bars indicate SD; n.s. = not significant. Data are the mean of three independent experiments.

5.4   Gliogenesis

5.4.1 Assessment of gliogenesis using Olig2 immunofluorescence

The shRNA driven knock-down of Fam72a in adult mouse NPC led to increased neurogenesis (Benayoun et al., 2014). This could either be the result of Fam72a increasing the maintenance of NPCs without affecting the fate towards neurons or of the reduced force of Fam72a towards a gliogenic fate of the NPCs. If option one were real, one would expect that the increased NPC maintenance is reflected in a higher number of Sox2+ cells, which was not the case (Fig. 20, 21 and 24). To examine the possible effect on gliogenesis, we decided to use oligodendrocyte transcription factor 2 (Olig2) as a marker protein. However, there were no significant differences in the percentage of RFP+ cells that are Olig2+ upon FAM72A or FAM72D expression (Fig. 26). These results suggest that neither FAM72A nor FAM72D significantly affected the gliogenesis in the developing medial mouse neocortex during late neurogenesis.

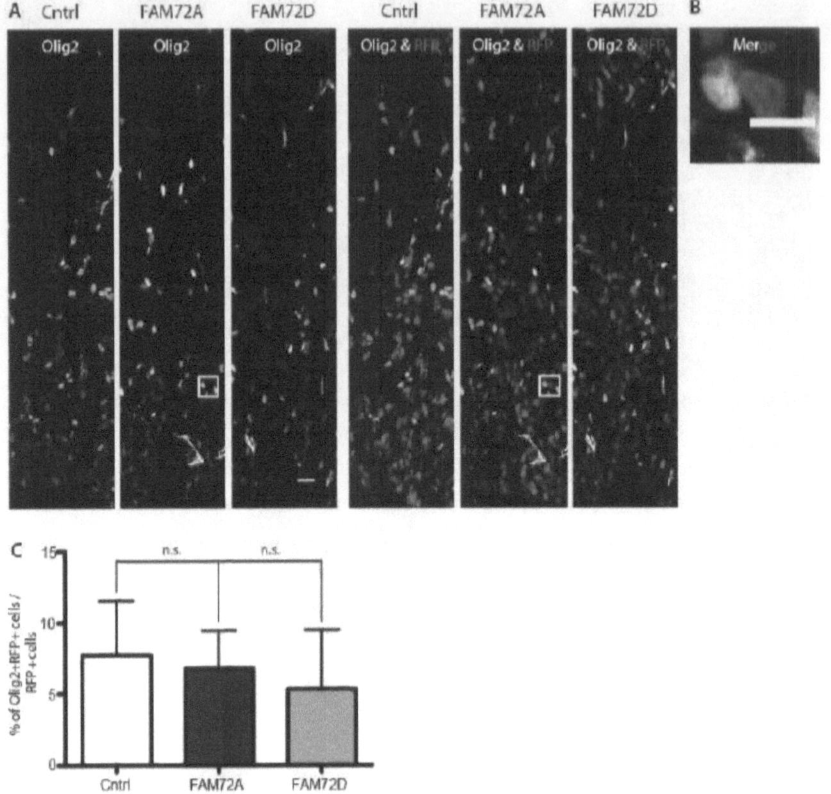

Fig. 26 – Assessment of gliogenesis using Olig2 immunofluorescence at late neurogenesis

Mouse neocortex was electroporated at E15.5 with the nuclear-targeting RFP-expressing plasmid either plus control (Cntrl), FAM72A- or FAM72D-expressing plasmids and analyzed at E18.5 (72 h post-IUE).

(A) Double immunofluorescence for Olig2 (grey) and RFP (red). Solid boxes indicate area in the SVZ-IZ that are shown at higher magnification in B (nucleus surrounded by dashed lines is Olig2+RFP+ double positive). Scale bar: 20 μm (A), 10 μm (B).

(C) Quantification of the percentage of total RFP+ cells that are Olig2+. There is no significant difference between control (Cntrl), FAM72A and FAM72D. Error bars indicate SD; n.s. = not significant. Data are the mean of four independent experiments.

5.5   Conclusion

It must be concluded that the forced expression of FAM72A or FAM72D in the developing mouse medial neocortex does not significantly affect the abundance of Sox2+ or Tbr2+ cells, the quantity of Ki67 + proliferating cells and the gliogenesis assessed by Olig2.

6 Results IV

6.1 Differences in gene expression upon ancestral FAM72A and human-specific FAM72D expression at mid-neurogenesis

6.1.1 Rationale and experimental setup

In this study, I aimed to functionally characterize the human-specific gene family *FAM72* in the embryonic mouse neocortex. Based on the findings of previous studies which suggest a potential role of the FAM72A and Fam72a protein in terms of cell cycle regulation (see: 1.6.3 and 7.3.1) in utero experiments were carried out to directly study effects on the abundance of NPCs and their proliferation in various developmental settings, such as different time points and locations within the embryonic mouse neocortex (see: Results II and III).

To get a broader idea of what the occurrence of FAM72A or FAM72D in NPCs of the developing neocortex could change in the machinery of the cells, we decided to perform a transcriptome analysis upon the expression of FAM72A or FAM72D in the embryonic mouse dorsolateral neocortex. Therefore, at least two E13.5 embryos of the same litter were in utero electroporated with an empty plasmid DNA vector (pCAGGS, control), a vector driving expression of FAM72A (pCAGGS-*FAM72A*) or *FAM72D* (pCAGGS-*FAM72D*) (>6 embryos in total per litter). Since we did not expect FAM72A or FAM72D to act as a transcription factor causing direct gene expression changes, we decided to harvest the embryos 24 h post-IUE at E14.5. The electroporated dorsolateral neocortical areas were microdissected, and GFP+ (electroporated) cells were isolated from the microdissected tissues by the fluorescence activated cell sorting (FACS) [4]. In total, four litters were independently processed. After RNA isolation, the RNA quality of each experiment was assessed using the RNA integrity number (RIN). Finally, three RNA samples of each condition were RNA sequenced [5]. Subsequently, the reads were checked for their overall quality and afterward aligned against the mouse reference genome (GRCm38) to quantify the expression of all expressed genes. Then, a differential gene expression analysis between the control, FAM72A and FAM72D samples on the raw counts was performed using a cutoff of $p < 0.01$. The resulting sets of differentially

---

[4] In collaboration with Ina Nüsslein, FACS Service Leader, MPI CBG

[5] In collaboration with Andreas Dahl and the Deep Sequencing group at BIOTEC (Dresden)

expressed genes were tested for enrichment in GO terms and pathways (KEGG, Reactome) on a 1% q-value level. [6, 7]

6.2 Differentially expressed genes upon ectopic FAM72A and FAM72D expression in the developing mouse dorsolateral neocortex

In total, there were 87 and 90 genes upregulated upon FAM72A and FAM72D expression, respectively. 15 out of these 177 genes showed a higher expression in the FAM72A and FAM72D expressing cells compared to the control cells (Table 1). Furthermore, the transcriptional analysis revealed 52 (ectopic FAM72A expression) and 67 (ectopic FAM72D expression) genes downregulated, 7 out of these 109 genes were shared by FAM72A and FAM72D (Fig. 27).

The mRNA of none of the marker proteins used to determine the parameters of neocortical development like NPC proliferation and abundance was found to be differentially expressed in our dataset. This additionally supports the findings assessed on the protein level using immunofluorescence and manual quantification presented in the Results II and III sections.

To better understand the functional implications of these findings, a systematic literature research was conducted to identify those genes which were considered to be potentially relevant for the physiological function of the FAM72A and FAM72D proteins in a cell.

---

[6] In collaboration with Holger Brandl, Senior Bioinformatics Data Engineer of the Scientific Computing Facility at MPI CBG / CSBD

[7] In collaboration with Marta Florio, Huttner Group MPI CBG, now: Harvard Medical School, Department of Genetics (Boston, United States)

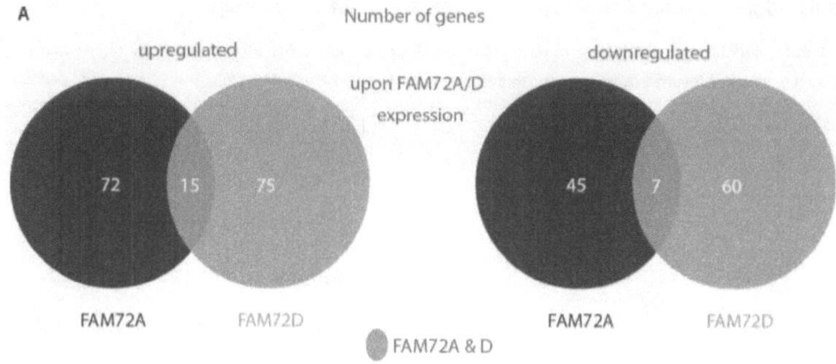

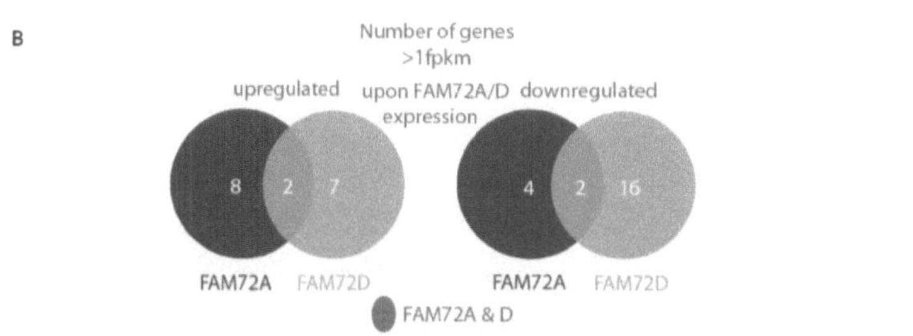

Fig. 27 – Differentially expressed genes upon ectopic FAM72A and FAM72D expression

Differentially expressed genes 24 h upon IUE of pCAGGS-FAM72A or -FAM72D plasmids in the developing dorsolateral mouse neocortex at mid-neurogenesis (E14.5) compared to IUE of empty pCAGGS plasmids (control) (A) The Venn diagram shows the numbers of up- or downregulated genes upon FAM72A or FAM72D expression in the developing mouse dorsolateral neocortex as well as the overlap between FAM72A and FAM72D compared to the control sample: e.g. 72 genes are only upregulated upon ectopic FAM72A, 15 upon FAM72A and FAM72D and 75 only upon FAM72D expression.

(B) The Venn diagram shows the number of up- or downregulated genes from the genes included in (A) that exhibited an expression >1 fpkm: e.g. 8 out of 72 genes upregulated upon ectopic expression of FAM72A.

6.3   Upregulated genes upon the ectopic FAM72A or FAM72D expression

6.3.1 Upregulated genes upon the ectopic FAM72A and FAM72D expression

First of all, I checked which genes were significantly upregulated upon the ectopic expression of FAM72A and FAM72D to recognize conserved expression patterns within the gene family.

Table 1: Upregulated genes upon the ectopic FAM72A and FAM72D expression

All genes found to be upregulated upon the ectopic expression of FAM72A and FAM72D. *Shisa5* and *Syde1* are highlighted in bold since they show an expression level >1 fpkm upon ectopic expression of FAM72A and FAM72D.

Short name	Full name
Aqp9	Aquaporin 9
Ephx4	Epoxide hydrolase 4
Matk	Megakaryocyte-associated tyrosine kinase
Paqr7	Progestin and adipoQ receptor family member VII
Ptgis	Prostaglandin I2 (prostacyclin) synthase
Rab43	RAB43, member RAS oncogene family
Shisa5	Shisa family member 5
Stk17b	Serine/threonine kinase 17b (apoptosis-inducing)
Syde1	Synapse defective 1, Rho GTPase, homolog 1
Thsd7a	Thrombospondin, type I, domain containing 7A
Tnfrsf13c	Tumor necrosis factor receptor superfamily, member 13c
Vegfc	Vascular endothelial growth factor C
Wdr72	WD repeat domain 72
Wfikkn1	WAP, FS, Ig, KU, and NTR-containing protein 1
Wnk4	WNK lysine deficient protein kinase 4

The identified genes exhibit a wide range of functions such as membrane water channel (Aqp9), steroid membrane receptor (Paqr7), vesicular transport (Rab43), endothelial cell migration (Thsd7a), B cell survival (Tnfrsf13c) and vasculo- and angiogenesis (Vegfc).

6.3.2 Upregulated genes upon the ectopic FAM72A or FAM72D expression – cut off: fpkm >1

Frequently only genes with expression higher than 1 fragment per kilobase of transcript per million mapped reads (fpkm) are further considered in transcriptome analyses. If this criterion is applied to our transcriptome dataset, it turns out that 10 out of the 87 upregulated genes upon FAM72A expression remain in the list after applying the >1 fpkm cut off regarding the gene expression level and 9 out of 90 in case of FAM72D (Fig. 27).

Table 2: Genes upregulated upon ectopic expression of FAM72A or FAM72D with an expression level >1 fpkm

The table shows all genes with an expression level > 1fpkm upregulated upon the ectopic FAM72A or FAM72D expression. *Shisa5* and *Syde1* are highlighted in bold. In contrast, the genes written in green are specifically upregulated upon the expression of the human-specific paralogue (*FAM72D*) (see: 6.4).

FAM72A >1 fpkm	FAM72D >1 fpkm
Hmga1	*Tapbp*
Cldn9	***Shisa5***
Shisa5	*Mtfp1*
Irf1	*Slitrk5*
Slc37a3	*Parp9*
Zfp383	***Syde1***
Fbxl5	*Cnp*
Syde1	*Rbm43*
Sult4a1	*Gm5689 (pseudogene)*
Fam161b	

6.3.3 Upregulated genes upon the ectopic FAM72A and FAM72D expression – cut off: fpkm >1

To better understand the general function of the *FAM72* gene family, I compared both lists (Table 2) and identified two genes upregulated upon FAM72A and FAM72D: *Shisa5 and Syde1*.

Shisa5 (putative NF-kappa-B-activating protein, Scotin) encodes for a protein localized to the endoplasmic reticulum (ER) and was shown to induce apoptosis together with p53 in a caspase-dependent manner (Bourdon et al., 2002; Terrinoni et al., 2004; Draeby et al., 2007). Furthermore, Shisa5 might be involved in the unfolded protein response (UPR)-induced apoptosis of the Latent Membrane Protein 1 (LMP1) Oncogene of Epstein-Barr virus (Pratt et al., 2012), which is particularly interesting in connection to the findings of Wang et al. (2011) that show a induced expression of FAM72A upon LMP1 transfection of 293 cells. Besides it is a new p63 target gene induced during epithelial differentiation (Zocchi et al., 2008).

Syde1, the synapse defective Rho GTPase homolog 1, was found to be a negative regulator of the endothelial barrier function (Amado-Azevedo et al., 2017). Besides, the expression of Syde1 was recently shown in human placentas and a direct effect on cytoskeletal remodeling, cell invasion and migration demonstrated. Furthermore, a reduced vascularization, as well as a breached barrier structure between maternal and fetal blood circulation, was found in Syde1-knockout placentas (Lo et al., 2017). Particularly interesting in a context of brain development, are results of experiments in mSYD1A knockout mice showing that the lack of the protein leads to a significantly reduced number of synaptic vesicles docking to the active zone and an impaired synaptic transmission, but no alteration in the general assembling of the synapses (Wentzel et al., 2013). Both genes could become particularly attractive if the evolutionary benefit of the expansion the family with sequence similarity 72 (FAM72) throughout human evolution is mainly due to an increased gene dosage or subfunctionalization (see: 1.3.3).

6.4   Upregulated genes specifically upon the ectopic FAM72D expression – cut off: fpkm >1

Regarding the second possibility that neofunctionalization occurred in case of the FAM72D protein, it would be important to consider those genes further that are specifically up- or downregulated upon the ectopic expression of the human-specific paralogue (Table 2, green). This section will focus on the genes specifically upregulated upon FAM72D expression: *Tapbp, Mtfp1, Slitrk5, Parp9, Cnp, Rbm43*, which were characterized using the existing literature.

6.4.1 Tapbp (TAP binding protein, Tapasin)

Tapbp (TAP binding protein, Tapasin) is an antigen processing molecule, which is part of the peptide-loading complex (PLC), a transient multi-subunit membrane complex in the endoplasmic reticulum, which releases stable peptide-MHC I complexes to the cell surface to provoke a T-cell response against malignant or infected cells. It consists of TAP (1 and 2), a transporter associated with antigen processing, ERp57, an oxidoreductase, the MHC-I heterodimer, and the endoplasmatic reticulum (ER) chaperones calreticulin (Crt) and Tapbp (Neefjes et al., 2011; Blees et al., 2017).

In fact, Tapbp mediates the interaction of the TAP transporter and newly assembled major histocompatibility complex (MHC) class I molecules and calreticulin. The critical functional role of Tapbp in the MHC class I-restricted antigen processing was established in experiments showing that expression of Tapbp in a negative mutant cell line (220) restores not only MHC class I-TAP association but also a regular MHC class I cell surface expression (Ortmann et al., 1997). Hence, the transient association of MHC class I molecules with Tapbp and TAP is crucial for the optimization of peptide cargo presented to CD8+ cytotoxic T cells (Momburg & Tan, 2002). In agreement with that, Tapbp was found to be significantly associated with tumor progression when downregulated in human melanoma lesions. Most likely, this alteration in the antigen-processing machinery leads to a failure of the acquired cellular immune system and thus, a lack of sufficient control of tumor progression and metastatic spread (Dissemond et al., 2003).

6.4.2 Mtfp1 (mitochondrial fission protein 1, Mtp18)

Mtfp1 (mitochondrial fission protein 1, Mtp18) is implicated in the maintenance of balanced fission and fusion of mitochondria in mammalian cells. Overexpression of the integral protein in the mitochondrial inner membrane results in fragmentation, whereas loss of Mtfp1 causes highly fused mitochondria (Tondera et al., 2004; Tondera et al., 2005; Wai & Langer, 2016). Interestingly, mitochondrial fission is blocked after overexpression of hFis1, another protein thought to be essential for the fission of mitochondria (Yoon et al., 2003), in cells with RNAi-meditated Mtfp1 knockdown suggesting that Mtfp1 is required for this process and indeed an essential contributor for the maintenance of mitochondrial morphology (Tondera et al., 2004; Tondera et al., 2005). In HaCaT (aneuploid immortal keratinocyte cell line) cells, reduced levels of Mtfp1 increase the sensitivity to apoptotic stimuli. Furthermore, it was shown that Mtfp1 expression is dependent on PI-3 kinase activity (Tondera et al., 2004). Besides that, Mtfp1 was found to act downstream of the mTORC1 signaling pathway critically regulating the phosphorylation of the essential fission GTPase dynamin-related protein (DRP1) (Tondera et al., 2005; Morita et al., 2017).
Suitably, both mitochondrial morphology and DRP1 phosphorylation were utterly reversed in asTORi-treated, TSC2 knockout and raptor knockout cells by the rescue of MTFP1 expression, which underscored the significant role of Mtfp1 in the control of mitochondrial dynamics (Morita et al., 2017). In addition, the critical part in the translational control of Mtfp1 by mTORC1/4E-BP was demonstrated by the increased apoptosis upon the uncoupling of Mtfp1 levels from the 4E-BP mediated regulation upon mTOR inhibition (Morita et al., 2017).

6.4.3 Slitrk5 (Slit and Ntrk-like protein 5)

Slitrk5 is part of the *Slitrk* (*Slit and Ntrk-like protein*) gene family, which are identified as integral membrane proteins containing two leucine-rich repeat (LRR) domains, similar to those of the axon guidance slit proteins and an intracellular C-terminal domain related to the sequence of trk (tyrosine kinases: TrkA, B and C) neurotrophin receptor proteins. *Slitrk1-5* were found to share 95-97% sequence homologies between mouse and human and to be predominantly expressed in the adult cerebral cortex in distinct but overlapping patterns (Beaubien & Cloutier, 2009; Takahashi & Craig, 2013) as well as in the human fetal brain. In cultured neuronal cells all Slitrk family members exhibited a neurite-modulating (Aruga et al., 2003) and synaptogenic activity through transsynaptic interactions of their ectodomains with presynaptic binding (Linhoff et al., 2009; Takahashi et al., 2012; Takahashi & Craig, 2013) hence, a range of functions which are known to be mediated by Trk receptors and their primary ligands like brain-derived neurotrophic factor (BDNF). Of the six CNS expressed Slitrk family members, Slitrk5 is the most exciting candidate to interact with the Trk receptor neurotrophin system, since Slitrk5 deficient mice showed neuronal phenotypes in the striatum comparable to those found in the brain of BDNF or TrkB knockout mice (Shmelkov et al., 2010; Song et al., 2015). Furthermore, experiments in *FLAG*-tagged TrkB and wt *Slitrk5* transfected HEK293 cells and in primary neurons with endogenous expression of both proteins demonstrated an interaction of Slitrk5 and TrkB mediated by the first LRR domain (LRR1) of Slitrk5. The interaction between Slitrk5 and TrkB seems to occur in a BDNF-dependent manner since BDNF stimulation significantly increased the co-localization of Slitrk5 and TrkB in striatal neurons as well as the interaction of both proteins in HEK293 cells. Furthermore, Slitrk5 is also capable (Takahashi et al., 2012) and even appears to primarily interact with the protein tyrosine phosphatase ∂ (PTP∂) under basal conditions, which is changed upon BDNF stimulation that directs Slitrk5 to interact with the TrkB receptor (Takahashi & Craig, 2013; Song et al., 2015). Song and colleagues could show that Slitrk5 knockout mice exhibit a significantly reduced steady-state TrkB receptor and TrkB downstream target activation as well as the requirement of Slitrk5 for an optimal long-term BDNF-dependent TrkB signaling in striatal neurons. Besides that, striatal neurons lacking Slitrk5 were characterized by an increased TrkB degradation, but could be rescued by the transfection of the wt Slitrk5, but not a chimeric Slitrk5 without the LRR1 domain. Finally, a positive effect of Slitrk5 on TrkB receptor recycling upon ligand treatment was identified as a cause for the enhanced degradation of TrkB in neurons lacking Slitrk5 (Song et al., 2015). Most likely, Slitrk5 plays a pivotal role in mediating the sorting of TrkB into Rab11-positive compartments after BDNF treatment, which was recognized to be important for the recycling of TrkB and the physiological function of TrkB signaling (Lazo et al., 2013; Song et al., 2015). Additionally, the Slitrk gene family was identified to provide candidate genes for several neuropsychiatric disorders (Takahashi & Craig, 2013). In case of Slitrk5, a link to the obsessive-compulsive-disorder (OCD) was found, since loss of Slitrk5 in mice did not only impair the corticostriatal synaptic transmission but also led to an excessive self-grooming and increased anxiety-like behavior – both typical symptoms of OCD, which were reduced by fluoxetine, a selective serotonin reuptake inhibitor (Shmelkov et al., 2010), commonly used to treat OCD in humans. This finding was further supported by the identification and functional characterization of Slitrk 5 mutations found in human patients suffering from OCD (Song et al., 2017).

6.4.4 Parp9 (Poly(ADP-ribose) polymerase 9)

Parp9 (Poly(ADP-ribose) polymerase 9) belongs to the family of Poly (ADP-ribose) polymerases (Parp) that is known to be involved in the regulation of various cellular processes such as DNA repair, genomic stability and programmed cell death. The activation of Parps was found to be one of the early responses to DNA damages (Herceg & Wang, 2001). Within the Parps, Parp9 seems to be different from other gene family members, since it lacks the Parp enzymatic activity (Zhang et al., 2015). Therefore, it hetero-dimerizes with Dtx3L (Deltex E3 Ubiquitin Ligase 3L) improving the antiviral response in mice and human cells treated with a modified form of the transcription factor Stat1. The Parp9-Dtx3l complex exhibits domains to interact with Stat1 and for the activity as an E3 ubiquitin ligase that indirectly promotes interferon-stimulated gene expression and facilitates the degradation of viral 3C proteases in the immunoproteasome. Interestingly, Parp9-Dtx3l were found to be able to auto-amplify upon the interferon response that developed upon infection (Zhang et al., 2015). Experiments in Parp9 knockout mice, however, revealed no evident phenotype suggesting that Parp9 may not be essential to achieve physiological double-stranded DNA breaks or the existence of strong compensatory mechanisms (Robert et al., 2017), additionally supporting a special role within the Parp gene family and a dependency in regard to its function on the hetero-dimerization with Dtx3l (Yang et al., 2017).

6.4.5 Cnp (2',3'-Cyclic-nucleotide 3'-phosphodiesterase)

Cnp (2',3'-Cyclic-nucleotide 3'-phosphodiesterase) is a membrane-associated enzyme within the myelin sheath of vertebrates (Kursula, 2006; Hinman et al., 2008). To date, the precise physiological function of Cnp remains vague especially regarding the conserved phosphodiesterase domain which is capable of hydrolyzing 2', 3' -cyclic nucleotides into 2' – nucleotides (Whitfeld et al., 1955). The Cnp activity was mostly associated with myelinated regions of the nervous tissue (Kurihara & Tsukada, 1967; Olafson et al., 2011). Though Cnp is not exclusively expressed in oligodendrocytes, the levels of expression in all other analyzed tissues to date are much lower (Toma et al., 2007; de Monasterio-Schrader et al., 2012; Myllykoski et al., 2016). Cnp expression was found in glioblast-like cells residing in the SVZ, during early stages of oligodendrocyte differentiation and remained to be expressed in mature oligodendrocytes (Braun et al., 1988; Yu, 1994). In mammals, two isoforms can be distinguished: Cnp1 and 2. In contrast to Cnp1, Cnp2 contains a mitochondrial targeting signal (Douglas et al., 1992; Monoh et al., 1993). Albeit, these extra residues are cleaved during the import resulting in an aa sequence similar to Cnp1 (Myllykoski et al., 2016).

Under pathological conditions like stroke or traumatic brain injury, the mitochondrial permeability transition pore (mPTP) is formed in the inner membrane upon the influx of calcium into the mitochondria and increases the mitochondrial membrane permeability which can cause mitochondrial swelling and finally the death of the cell (Lemasters et al., 2009). The mitochondrial Cnp was found to be a regulator of the calcium threshold for the mPTP (Azarashvili et al., 2009) for example age-dependent in brain mitochondria of rats showing a lower calcium threshold in mitochondria from old rats with reduced Cnp levels compared to those of young rats (Krestinina et al., 2015). These findings become especially interesting considering that the prolonged opening of the mPTP can be involved in neurodegenerative diseases due to the caused apoptosis or necrosis (Rasola & Bernardi, 2011, 2014). Conversely to Cnp the level of its substrate 2',3'-cAMP correlates positively with the opening of the mPTP (Azarashvili et al., 2009) hence, Cnp seems to have a protective function in the context of brain injury, since it participates in the degradation pathway of 2',3'-cAMP to adenosine (Verrier et al., 2013) which acts protective towards neurons (Stone et al., 2007). Thus, oligodendrocytes, where the 2',3'-cAMP to adenosine pathway due to Cnp has its highest expression, may protect axons from lasting injury (Verrier et al., 2013). Another observation of Cnp is its interaction with microtubules, and the ability to promote tubulin polymerization (Bifulco et al., 2002) mediates process outgrowth in oligodendrocytes (Lee et al., 2005). Albeit, Cnp does not seem to be essential for the correct ultrastructure and physical stability of the axonal myelin, but its absence in mice brains leads to axonal swellings and neurodegeneration throughout the brain causing hydrocephalus and premature death (Lappe-Siefke et al., 2003).

The protective function of Cnp is additionally underpinned by findings showing that artificial silencing of Cnp expression leads to increased levels of inflammatory mediators (Myllykoski et al., 2016). Furthermore, higher levels of Cnp are found in activated microglia cells (L. Yang et al., 2014) probably causing a reduced production of inflammatory mediators in activated microglia cells via the 2',3'-cAMP-adenosine pathway (Newell et al., 2015). Neonatal oxytocin treatment in rats, however, reduced Cnp mRNA in contrast to the increased expression of neuron-specific enolase (NSE) and the glial fibrillary acid protein (GFAP) (Havranek et al., 2017). Moreover, altered Cnp levels were linked to psychiatric diseases such as schizophrenia or depression (Peirce et al., 2006; Hagemeyer et al., 2012; Rajkowska et al., 2015; Myllykoski et al., 2016). Cnp was identified to be the most robustly reduced candidate among the oligodendrocyte-associated genes on the mRNA and protein level in brains of schizophrenic (Mitkus et al., 2008), major depressive (Aston et al., 2005) or bipolar (Tkachev et al., 2003) patients suggesting a critical role rather for a more general pathology than to be restricted to one specific diagnosis. Hagemeyer et al. (2012) further identified the reduced expression as a primary cause for a behavioral phenotype that only occurred upon aging as an additional 'pro-inflammatory hit' and manifested in white matter abnormalities and neurodegeneration in mice as well as humans.

6.4.6 Rbm43 (RNA binding motif protein 43)

In the case of **Rbm43** (RNA binding motif protein 43), neither *Pubmed* nor *Google Scholar* provided any useful published data on this gene to date (March 2019). This is why I focused on GO – annotations (www.geneontology.org) to roughly determine the molecular function of the protein. The result is shown in Table 3.

Table 3: Upregulated genes specifically upon the ectopic FAM72D expression – cut off: >1 fpkm

Brief characterization of the genes only upregulated upon FAM72D expression but not FAM72A expression with an expression level >1 fpkm.

Gene name	Comment
Tapbp *(TAP binding protein)*	Maturation of MHC class I molecules in the endoplasmatic reticulum (Ortmann et al., 1997; Momburg & Tan, 2002; Yinan Zhang & Williams, 2006; Blees et al., 2017).
Mtfp1 *(Mitochondrial fission process 1, Mtp18)*	Maintenance of mitochondrial morphology by contributing to the control of mitochondrial fission (Tondera et al., 2005; Morita et al., 2017)
Slitrk5 *(SLIT and NTRK-like family, member 5)*	Slitrks: neurite outgrowth, synaptogenesis, and neuronal survival (Aruga & Mikoshiba, 2003; Proenca et al., 2011; Song et al., 2015) Slitrk5: TrkB co-receptor mediating its BDNF dependent trafficking and signaling (Song et al., 2015)
Parp9 *(Poly (ADP-ribose) polymerase family, 9)*	Parps: DNA repair, genomic stability, programmed cell death (Herceg & Wang, 2001) Parp9: hetero-dimer with Dtx3l improves the antiviral response (Zhang et al., 2015)
Cnp *(2',3'-cyclic nucleotide 3' phosphodiesterase, CNPase)*	Anti-inflammatory, neuro-protective, oligodendrocyte-associated protein (Myllykoski et al., 2016)
Rbm43 *(RNA binding motif protein 43)*	GO – molecular function: *RNA binding* (GO:0003723, IEA)

6.5   Downregulated genes upon the ectopic FAM72A or FAM72D expression

Not only increased levels of expression of a gene can be of functional relevance. It is also reasonable to assume that the downregulation of a gene involved in the regulation of a certain biological process can be responsible for a phenotype of interest.

6.5.1 Downregulated genes upon the ectopic FAM72A or FAM72D expression – cut off: fpkm >1

Of the 52 genes downregulated upon FAM72A expression, six remained in the list when I focused only on the genes with an expression level of more than 1 fpkm in the control sample (genes with an expression <1 fpkm due to the downregulation upon FAM72A expression were included). Expression of FAM72D led to 67 downregulated genes out of which 18 were expressed with more than 1 fpkm in the control. Like in the case of the upregulated genes, there were two genes found to be downregulated upon the ectopic expression of FAM72A and FAM72D: *Olfr543 and Galnt4.*

6.5.2 Downregulated genes upon ectopic FAM72A expression – cut off: fpkm >1

The genes downregulated in the brains with ectopic FAM72A expression are listed in Table 4 including a brief description of known functions of these genes. Potentially most interesting might be the *Spermidine N1-acetyltransferase 1 (Sat1)* which was found to be a blood biomarker for suicidality, since it directly connects the molecular to the behavioral dimension. Likewise, it is obvious to consider a tumor-associated gene like *Galnt4* in the context of a study with a particular interest in cell proliferation.

Table 4: Genes downregulated upon FAM72A with an expression level >1 fpkm in the control sample

The genes downregulated upon FAM72A **and** FAM72D with an expression level >1 fpkm in the control sample are highlighted in bold.

Gene name	Comment
Sat1 *(Spermidine/spermine N1–acetyltransferase 1)*	Blood biomarker for suicidality (Le-Niculescu et al., 2013; Niculescu et al., 2017)
Gm6525	Predicted pseudogene 6525
Olfr543 *(Olfactory receptor 543)*	Olfactory receptor (Ensembl ID: ENSMUSG00000044814)

Hba-x	GO – biological process: erythrocyte maturation
(Hemoglobin X, alpha-like embryonic chain in Hba complex)	(GO:0043249, IEA), oxygen transport (GO:001567, IEA)
Tfe3	Transcription factor (Mansky et al., 2002;
(Transcription factor E3)	Steingrimsson et al., 2002)
Galnt4	N-acetylgalactosaminyl transferase, tumorigenic
(Polypeptide N-Acetylgalactosaminyltransferase 4)	(Zhang J et al., 2016; Qu et al., 2017)

(Table 4 continued)

6.5.3 Downregulated genes upon ectopic FAM72D expression – cut off: fpkm >1

The resulting list of genes which fulfill these parameters is composed of several interesting genes including *Dusp1,* which was found to mediate cell proliferation as well as the cancer development and *Gsg2*. But also, *Arx* and *Tsnax*, two genes associated with neurological and psychiatric diseases and others are listed in Table 5.

Table 5: Genes downregulated upon FAM72D with an expression level >1 fpkm in the control sample

The genes downregulated upon FAM72A **and** FAM72D with an expression level >1 fpkm in the control sample are highlighted in bold (*Olfr543, Galnt4*) and the remaining specifically downregulated genes upon FAM72D expression in green.

Gene name	Comment
Rsrp1 *(Arginine and Serine Rich Protein 1)*	GO – molecular function: protein binding (GO:0005515, IPI)
Med11 *(Mediator complex subunit 11)*	GO – molecular function: RNA polymerase II transcription cofactor activity (GO:0001104, IEA), Ubiquitin protein ligase activity (GO:0061630, IEA), part of the mediator head module (Seizl et al., 2011)
Dusp1 *(Dual specificity protein phosphatase 1)*	Inhibition of cell proliferation, metastasis and invasion and angiogenesis in gallbladder cancer (Shen et al., 2017), Mapk pathway regulation (Kang et al., 2016; Lopes et al., 2017)
Bcdin3d *(BCDIN3 Domain Containing RNA Methyltransferase)*	RNA methyltransferase involved in miRNA processing and breast cancer progression

	(Xhemalce et al., 2012; Yao et al., 2016; Martinez et al., 2017)
Ptx3 *(Pentraxin-related protein PTX3, TNF-inducible gene 14 protein, TSG-14)*	Involved in innate immunity, inflammation, matrix deposition (Garlanda et al., 2006)
Gsg2 *(Histone H3 Associated Protein Kinase, Haspin)*	Cell cycle progression (Nguyen et al., 2014; Quartuccio et al., 2017)
Mpnd *(MPN domain containing)*	GO annotation: peptidase activity (GO:0008233, IEA), hydrolase activity (GO:0016787, IEA)
Hgh1 *(Human growth hormone 1)*	Human growth hormone homolog (Ensembl ID: ENSMUSG00000022554)
Ankrd16 *(Ankyrin repeat domain 16)*	SNPs associated with breast cancer subtypes (O'Brien et al., 2014)
Olfr543 *(Olfactory receptor 543)*	Olfactory receptor (Ensembl ID: ENSMUSG00000044814)
Bloc1s4 *(Biogenesis of lysosomal organelles complex-1, subunit 4, cappuccino)*	Subunit of the biogenesis of lysosome-related organelle complex-1 (Bloc1), which mediates trafficking at the endosome (Huang et al., 2012), associated with Hermansky-Pudlak syndrome (Ciciotte et al., 2003; Li et al., 2003)
Fkbp7 *(FK506 binding protein 7, Fkbp 23)*	Peptidyl-prolyl isomerase with EF-hand motif localized to the endoplasmic reticulum (Boudko et al., 2014)
Arx *(Aristaless related homeobox)*	Mutations associated to structural (Coman et al., 2017) and functional neurodevelopmental disorders, positively regulates Wnt / β-catenin signaling (Cho et al., 2017)
Kctd6 *(Potassium channel tetramerization domain containing 6)*	Cullin-dependent regulation of small ankyrin-1 protein turnover (Lange et al., 2012)
Siah1b *(Seven in absentia 1B)*	Control of cerebellar granule neuron's germinal zone exit (Famulski et al., 2010)
Tsnax *(Translin-associated factor X, trax)*	DNA damage response, cell proliferation control, genetic association to psychiatric diseases (Jaendling & McFarlane, 2010; Duff et al., 2013)
Galnt4 *(Polypeptide N-Acetylgalactosaminyltransferase 4)*	N-acetylgalactosaminyl transferase,

	tumorigenic activity
	(Zhang J et al., 2016; Qu et al., 2017)
Icmt *(Isoprenylcysteine carboxyl methyltransferase)*	Modification of C-terminal CaaX motifs, Regulation of NOTCH signaling by RAB7 and RAB8 (Court et al., 2017)

6.6 Genes previously shown to be differentially expressed upon forced FAM72A expression

6.6.1 Cell cycle regulators

Besides, it was asked whether cell cycle regulators shown to be differentially expressed in Western Blot analyses upon FAM72A expression in a nasopharyngeal tumor cell line (Wang et al., 2011) would also be differentially expressed in our dataset. None of the proteins found to be up- or downregulated was identified as such on the mRNA level neither upon FAM72A nor FAM72D ectopic expression in the developing mouse neocortex at mid-neurogenesis.

Table 6: Cell cycle regulators previously shown to be differentially expressed upon forced FAM72A expression

None of the cell cycle regulators shown to be differentially expressed upon FAM72A in Wang et al. (2011) is up- or downregulated upon expression FAM72A or FAM72D in the developing mouse neocortex.

Gene name	Differentially expressed
Cyclin D1	no
E2F1	no
CDK2	no
CDK4	no

(Table 5 continued)

6.6.2 Tumor suppressor genes

Wang et al. (2011) identified four tumor suppressor genes up- (*p16, p21, p53*) or downregulated (*p19*) upon the overexpression of FAM72A.

None of them was differentially expressed in our transcriptome data suggesting that the cellular pathways the FAM72 proteins are involved in depend on the biological setting such as the cell type.

6.6.3 Proteins previously observed to interact with FAM72A

Using a yeast two-hybrid screening test, five proteins interacting with FAM72A [8] : TMEM115/PL6, YPEL3, ERP44, CDK5RAP3, and NNAT, were previously identified (Heese, 2013). The latter one, neuronatin (NNAT) is mainly expressed in the brain during neurogenesis and assumed to affect neuronal growth as well as the differentiation of pluripotent stem cells into a neural fate. Additionally, it tends to aggregate causing pathologies also outside the brain such as Lafora disease and diabetes (Joseph, 2014). We used our transcriptome dataset to determine whether one of the mouse orthologues of these proteins is differently expressed on the mRNA level upon the expression of its putative interactor, but that was not the case for any of them as well as for the DNA glycosylase 2 (UNG2) as another interactor of FAM72A identified via a pull-down assay (Guo et al., 2008).

6.7 Effect of ectopic FAM72A and FAM72D expression on genes implicated in neural lineage fate decision [9]

To address this question, we asked how many of the genes differentially expressed upon FAM72A or FAM72D were found to be enriched in NPCs or neurons in the mRNA mouse expression dataset from Florio et al. (2015) at E14.5 (same stage as our dataset). Interestingly, the analysis showed that there are more genes found to be enriched in NPCs which are upregulated upon the expression of FAM72D compared to FAM72A and more genes enriched in neurons that are upregulated upon FAM72A compared to FAM72D expression (Fig. 28). In total 11 genes are enriched in NPCs and upregulated upon FAM72D, 5 (*Mtfp1, Cnp, Parp9, Tapbp, Rbm43*) out of these (Table 3, green) are specifically upregulated upon FAM72D but not FAM72A and survived the >1 fpkm expression level cut off. Interestingly, *Syde1* also belongs to the NPC enriched genes, though to date it was only found to be involved in neuronal processes such as synaptic transmission (Wentzel et al., 2013).

In the case of the downregulated genes, it is the other way around: there are more NPC enriched genes downregulated upon FAM72A and more neuron-enriched genes downregulated upon FAM72D compared to FAM72A expression (Fig. 28). This finding slightly

---

[8] Though the author (Heese, 2013) claims to use a FAM72A (p17) construct in this study, he refers to isoform 2 in Nehar et al., 2009 as a reference sequence. Albeit, Fig. 1C in Nehar et al., 2009 shows an Isoform-2 amino acid sequence that is identical to the FAM72B reference sequence (ENSG00000188610).

[9] In collaboration with Marta Florio, Huttner Group MPI CBG, now: Harvard Medical School, Department of Genetics (Boston, United States)

supports the idea that FAM72D rather than FAM72A could positively affect the fate towards the maintenance of NPC characteristics.

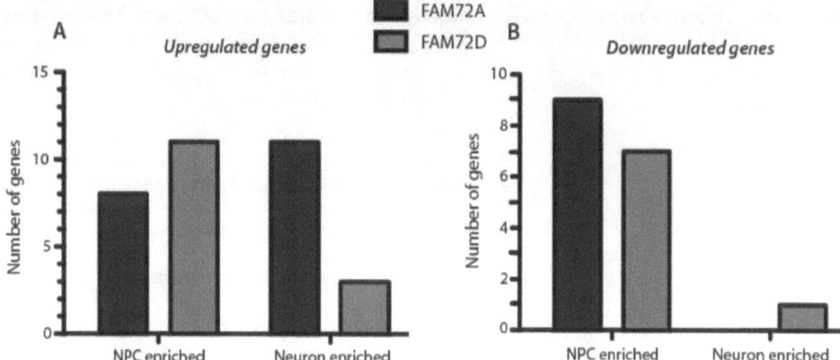

Fig. 28 – Genes differentially expressed upon ectopic FAM72A or FAM72D expression found to be enriched in NPCs or neurons in Florio et al. (2015)

(A) There are more genes found to be enriched in NPCs which are upregulated upon the expression of FAM72D (11 genes) compared to FAM72A expression (8 genes) and more genes enriched in neurons that are upregulated upon FAM72A (11 genes) compared to FAM72D expression (3 genes).
(B) There are more NPC enriched genes downregulated upon FAM72A (9 genes) compared to FAM72D expression (7 genes) and no neuron enriched gene downregulated upon FAM72A, but 1 upon FAM72D expression.

The following Tables were created to give an overview of the variety of genes differentially expressed upon FAM72A and FAM72D ectopic expression and to prevent a potentially misleading filter caused by the >1 fpkm expression level cut off mostly used for the previous analyses. Furthermore, they provide interesting insights about the enrichment of genes differentially expressed upon the expression of a human-specific gene and its ancestral paralogue in mouse NPC and neurons, during mid-neurogenesis, a critical period for neocortical development.

6.7.1 Upregulated and NPC-enriched genes

Tables 7 and 8 show all NPC enriched genes upregulated upon ectopic FAM72D expression and all NPC enriched genes upregulated upon ectopic FAM72A expression according to Florio et al. (2015).

Table 7: Genes *upregulated* upon ectopic *FAM72D* expression and enriched in *NPCs* according to Florio et al. (2015)

The five genes that were only upregulated upon FAM72D but not A with an expression >1 fpkm are highlighted in green. *Syde1* is highlighted in bold because it is upregulated upon FAM72A and FAM72D and shows an expression >1 fpkm.

Short name	Full name
Mtfp1	Mitochondrial fission process 1
Cnp	2',3'-cyclic nucleotide 3' phosphodiesterase
L2hgdh	L-2-hydroxyglutarate dehydrogenase
Ror2	Receptor tyrosine kinase-like orphan receptor 2
Dct	Dopachrome tautomerase
Parp9	Poly (ADP-ribose) polymerase family, member 9
Tapbp	TAP binding protein
Stk17b	Serine/threonine kinase 17b
Nlrx1	NLR family member X1
Syde1	Synapse defective 1, Rho GTPase, homolog 1
Rbm43	RNA binding motif protein 43

Table 8: Genes *upregulated* upon ectopic F*AM72A* expression and enriched in *NPCs* according to Florio et al. (2015)

Syde1 is highlighted in bold because it is one of the two genes upregulated upon FAM72A and FAM72D expression and shows an expression >1 fpkm.

Short name	Full name
Irf1	Interferon regulatory factor 1
Stk17b	Serine/threonine kinase 17b (apoptosis-inducing)
Tyro3	TYRO3 protein tyrosine kinase 3
Efna1	Ephrin A1
Col4a5	Collagen, type IV, alpha 5
Syde1	Synapse defective 1, Rho GTPase, homolog 1 (C. elegans)
Hmga1	High mobility group AT-hook 1
Cldn9	Claudin 9

6.7.2 Downregulated and NPC-enriched genes

Table 9 shows all NPC enriched genes according to Florio et al. (2015) downregulated upon ectopic FAM72D expression.

Table 9: Genes *downregulated* upon ectopic *FAM72D* expression and enriched in *NPCs* according to Florio et al. (2015)

Short name	Full name
4930452B06Rik	*RIKEN cDNA 4930452B06 gene*
Nfatc2	*Nuclear factor of activated T cells, cytoplasmic, calcineurin-dependent 2*
Plekha5	*Pleckstrin homology domain containing, family A member 5*
Grtp1	*GH regulated TBC protein 1*
Siah1b	*Seven in absentia 1B*
Gsg2	*Germ cell-specific gene 2*
Pgbd1	*PiggyBac transposable element derived 1*

Table 10 shows all NPC enriched genes according to Florio et al. (2015) downregulated upon ectopic FAM72D expression.

Table 10: Genes *downregulated* upon ectopic *FAM72A* expression and enriched in *NPCs* in Florio et al. (2015)

Short name	Full name
Lcp1	*Lymphocyte cytosolic protein 1*
Pla2g7	*Phospholipase A2, group VII (platelet-activating factor acetylhydrolase, plasma)*
Sat1	*Spermidine/spermine N1-acetyl transferase 1*
Sgk3	*Serum/glucocorticoid regulated kinase 3*
Aldh1l1	*Aldehyde dehydrogenase 1 family, member L1*
Nqo2	*NAD(P)H dehydrogenase, quinone 2*
Entpd1	*Ectonucleoside triphosphate diphosphohydrolase 1*
Prokr2	*Prokineticin receptor 2*
Pld4	*Phospholipase D family, member 4*

6.7.3 Upregulated and neuron-enriched genes

Table 11 shows all neuron-enriched genes according to Florio et al. (2015) upregulated upon ectopic FAM72D expression.

Table 11: Genes *upregulated* upon *FAM72D* and enriched in *neurons* according to Florio et al. (2015)

Shisa 5 is highlighted in bold because it is one of the two genes >1 fpkm upregulated upon FAM72A and FAM72D expression.

Short name	Full name
Shisa5	Shisa homolog 5
Kif26a	Kinesin family member 26A
Sec24d	Sec24 related gene family, member D

Table 12 shows all neuron-enriched genes according to Florio et al. (2015) upregulated upon ectopic FAM72A expression.

Table 12: Genes *upregulated* upon *FAM72A* and enriched in *neurons* according to Florio et al. (2015)

Short name	Full name
Grik3	Glutamate receptor, ionotropic, kainate 3
Lpl	Lipoprotein lipase
Gabrg2	Gamma-aminobutyric acid (GABA) A receptor, subunit gamma 2
Shisa5	Shisa homolog 5 (Xenopus laevis)
Lmo3	LIM domain only 3
Cdh13	Cadherin 13
Mgll	Monoglyceride lipase
Syt1	Synaptotagmin I
Sh3bp4	SH3-domain binding protein 4
Gsg1l	GSG1-like
Magel2	Melanoma antigen, family L, 2

6.7.4 Downregulated and neuron-enriched genes

Table 13 shows that there is only one neuron-enriched gene according to Florio et al. (2015) downregulated upon ectopic FAM72D expression.

Table 13: Genes *downregulated* upon *FAM72D* and enriched in *neurons* according to Florio et al. (2015)

Short name	Full name
Gnal	*Guanine nucleotide binding protein, alpha stimulating, olfactory type*

Table 14 shows that there is no neuron-enriched gene according to Florio et al. (2015) downregulated upon ectopic FAM72D expression.

Table 14: Genes *downregulated* upon *FAM72A* and enriched in *neurons* according to Florio et al. (2015)

Short name	Full name
-	-

6.8 GO enrichment analysis [10]

Finally, we used gene ontology (GO) to perform an enrichment analysis on the genes in our dataset to identify the GO terms that are over-represented using annotations for that gene set. This analysis only revealed the GO term angiogenesis to be enriched upon ectopic expression of FAM72A. The genes associated with this GO term are listed in Table 15.

---

[10] In collaboration with Marta Florio, Huttner Group MPI CBG, now: Harvard Medical School, Department of Genetics (Boston, United States)

Table 15: Genes enriched for the GO term angiogenesis found to be upregulated upon ectopic expression of FAM72A

The genes written in blue highlight those which are also upregulated in the FAM72D sample.

Short name	Full name
Efna1	*Ephrin A1*
Col18a1	*Collagen, type XVIII, alpha 1*
Vegf-c	*Vascular endothelial growth factor c*
Cdh13	*Cadherin 13*
Adam8	*A disintegrin and metallopeptidase domain 8*
Thsd7a	*Thrombospondin, type I, domain containing 7A*
Cx3cl1	*Chemokine (C-X3-C motif) ligand 1*
Htatip2	*HIV-1 Tat interactive protein 2*

Two of the eight genes in the cluster were also upregulated upon FAM72D expression: *Vegf-c* and *Thsd7a*.

Vegf-c belongs to the family of vascular endothelial growth factors which are known to affect the growth of blood vessels as well as neural cells (Raab & Plate, 2007). Vegf-c was found to be essential for the homeostasis and the initial steps in lymphatic development with high importance for the delamination of lymphatic progenitor cells from embryonic veins, for instance (Karkkainen et al., 2004; Raab & Plate, 2007). Like Vegf-a, Vegf-c increases vascular permeability and stimulates migration and mitosis of endothelial cells (Joukov et al., 1996; Lee et al., 1996). Mice lacking Vegf-c die before birth (Karkkainen et al., 2004), show a strongly decreased proliferation of Vegfr-3 expressing NPCs without intracerebral blood vessel defects and a selective loss of oligodendrocyte precursor cells in the embryonic optic nerve (Le Bras et al., 2006). Furthermore, Vegf-c via interaction with Vegfr-3 might function as a chemoattractant for microglia and macrophages (Shin et al., 2008; Shin et al., 2010).

Thsd7a was identified as endothelial cell migration and tube formation mediating protein that might be involved in cytoskeletal organization. Overexpression of a Thsd7a C-terminal fragment inhibited migration and disturbed tube formation of human umbilical vein endothelial cells, where it was found to be expressed at the leading edge during migration. Immunohistological analyses additionally revealed its co-localization with $a_v\beta_3$ integrin and paxillin (Wang et al., 2010).

6.9. Conclusion

The analysis of the data presented in Results IV provides evidence for a neofunctionalization of *FAM72D* compared to *FAM72A* primarily because of the surprisingly low overlap of genes differentially expressed upon in utero electroporation of *FAM72A* or *FAM72D* (see: 6.2 and 6.7). Beyond various interesting candidate genes that were identified in this screening, it is tempting to highlight the upregulation of Slitrk5 and Mtfp1 upon FAM72D in utero electroporation. Possible implications of these findings are discussed in chapter 7.3.

7   Discussion

7.1   What makes us human?

Albeit, "What makes us human?" was the primary underlying question of this study, a broad discussion including not only a great variety of scientific but also philosophical aspects would surely exceed the scope of this book. Thus, the following chapter will only briefly discuss a selection of some essential facets to facilitate the connection of the insights from the basic research presented in the previous sections and the major question about the human identity.

7.1.1 Bipedalism

One general concept already suggested by Charles Darwin states that the development of bipedalism was the essential process that made humans distinct from other mammals. Bipedalism favored additional capabilities such as tool making and use with the hands, which later selected for bigger brains and its cognitive skills such as a syntactical-grammatical language, autobiographical memory, symbolic thought, self-reflection, and the enormous social power. Hence, humans could become a successful species on earth despite their lack of athletic power (Lieberman, 2014; Lieberman, 2016).

7.1.2 The human brain

Today the brain is viewed as the core component of human identity. Maybe bipedalism was necessary for the development of an advanced mind, but without the evolved human brain the world we live in today would be unimaginable.

7.1.2.1 Qualitative leap vs. quantitative advancement

For a long time, humans were pretty much convinced that our cognitive and behavioral capacities would be unique, not only in degree but also in kind. It remains an intensively discussed question whether the distinctive skills of humans are the result of a qualitative leap or a quantitative advancement. However, more and more cognitive qualities like the theory of mind that was historically thought to be such a leap in human evolution turn out to rudimentary exist also in other apes (Krupenye et al., 2016; Sousa et al., 2017). The knowledge about the existence of these capabilities in other species is important to ask the right questions and to design proper experiments to unveil the relevant functional units within the brain constituting the outstanding characteristics of humanity.

7.1.2.2 Brain size and number of neurons

In any case, one tempting key parameter to consider is the increase of absolute brain size during primate and particularly human evolution. Though species such as elephants and whales have bigger brains than humans, we would vigorously defend the statement that human cognitive abilities exceed those of these species (Dunbar & Shultz, 2007; Sousa et al., 2017).

Based on the idea that bigger bodies need bigger brains to be controlled, the concept of *brain-to-body weight ratios* occurred. Indeed, in contrast to blue whales with a brain-to-body-weight ratio of about 0,01%, humans show a higher one with about 2%. But this concept lost power because smaller animals often show a brain-to-body-weight ratio even higher than 2% (Harvey et al., 1980). As a result, dozens of analyses and comparisons of brain sizes could not provide a robust model of how absolute brain size and intellectual capabilities correlate when various lineages spanning datasets are regarded. Within specific biological lineages some concepts such as the *encephalization quotient* suggest that the human brain is about 4-7 times bigger than it would be expected for a mammal of its size (Marino, 1998; Herculano-Houzel, 2009). The absolute brain size was found to be a good predictor for cognitive abilities within non-human primates (Deaner et al., 2007). Interestingly, species with the largest brains were also found to show a greater behavioral range and flexibility than those with smaller brains among mammals (Jerison, 1985; Marino, 2002) as well as among birds (Lefebvre et al., 2004; Herculano-Houzel, 2009).

Another way to look at the human brain is to regard its biological building blocks: mainly neurons and glial cells and indeed, the number of cortical neurons was found to correlate with intelligence (Roth & Dicke, 2005). But also this approach reveals the natural complexity of the issue, since the human cerebral cortex may consist of about 16 billion neurons (Azevedo et al., 2009; Herculano-Houzel, 2009), but the cerebral cortex of the long-finned pilot whale, for instance, was found to contain about twice as many (Mortensen et al., 2014). Additionally, it is essential to be aware that many of the analyses to date were performed based on insufficient data regarding the number of cortical neurons due to technical and methodological limitations of the experiments.

In the end, it is not surprising that the key to human cognition cannot be conclusively explained based on rather primitive mono-dimensional models such as the increase in absolute brain size or the neuronal number. Nevertheless, the rationale behind the idea that a bigger brain consisting of more neurons would possess a higher computational capacity intuitively makes sense and might be a good starting point to develop a more complex multi-dimensional concept.

7.1.2.3 Expansion of specific cortical regions

Insights about unique human abilities and the associated parts of the neocortex additionally made clear that the relative expansion of certain neocortical areas and networks has to be taken into consideration to explain the rapid development of human cognitive abilities within the last 3 million years. Not all of the brain's functions and regions expanded during this period, especially the primary sensory cortices are almost equivalent between chimpanzees and humans (Diamond & Hall, 1969). Consequently, the proportion of association cortices increased with the overall expansion of the neocortex and provided the capacity for cognitive functions that we consider to be *higher*. One following question is whether the connectivity patterns of these expanded regions maintain ancient or if they created new circuit properties. This leads back to the question whether human cognitive faculties arose from a stronger computational power (quantitative advancement) or if they got the needed space to newly develop (qualitative leap) through an increased neocortical size (Buckner & Krienen, 2013; Buckner & Krienen, 2017).

7.1.2.4 Insights from human brain pathologies

The importance of cortical size for normal brain function is clearly shown in clinical cases of patients with micro- or macrocephaly (small or enlarged brains) which show a variety of cognitive deficits (Sun & Hevner, 2014). At the same time, macrocephalic people usually are not smarter than the average, but rather show cognitive limitations most likely due to an imbalanced cortical growth. Besides that, some of the microcephalic patients do not have bigger brains than chimpanzees but are considered to be smarter. This may be in part explained by the human-specific structural organization of the brain as well as in the stronger cultural support facilitating the development of higher cognitive functions.

7.1.3 Hypotheses for enlarged brains

Generally, it is likely that the evolutionary development of the human brain occurred in close interaction with the environment and other parts of our ancestor's bodies. Bipedalism offered the opportunity to discover novel activities with our hands directed by the expanded association cortices, jaw muscles had to adapt to reduce physical restrictions for brain growth (Stedman et al., 2004) and many other adaptations occurred in parallel or advance to establish the brain we are used to thinking with today.

One of the most popular hypotheses to explain how the human brain increased in size is based on the *expensive tissue hypothesis* and highlights the importance of a changed diet of modern humans and adaptions in our digestive tract (Aiello, 1997). In this context, it has been proposed that the use of fire to cook meals in Homo erectus overcame the limitation of the available number of hours for feeding and the necessary caloric intake to provide enough energy for an increased neuronal number. The hypothesis furthermore suggests that the shift to a cooked diet was a positive driving force for human evolution since it increased the not-eating time and the capacities to create new tools and further improvements using the increase in computational power in our brain (Fonseca-Azevedo & Herculano-Houzel, 2012). In fact, this concept which seems to be reasonable in the first place is also quite controversial in the scientific community and the expensive tissue hypothesis regarding brain size expansion itself strongly questioned (Navarrete et al., 2011) as well as the concept that human brain expansion was dependent on the use of fire for cooking (Cornélio et al., 2016).

A different logical starting point is taken by the *cognitive buffer hypothesis*, which is based on the observation that larger brains tend to increase the flexibility to respond to unusual or complex socioecological challenges making increased survival rates possible, enabling a longer reproductive life and thereby compensated for a delayed reproduction, which also occurs due to a more extended period of maturation (Sol, 2009). A great variety of additional concepts (Dunbar & Shultz, 2017) such as the *social brain hypothesis*, that can be briefly described as the idea that large brains evolved to manage growing complex social systems (Dunbar, 2009), which are as controversial as the concepts introduced above (Barrett et al., 2007) unfortunately cannot be discussed here due to the scope of the book.

What we learn from this debate is that the causalities of why and how the brain enlarged during human evolution is still debated, but what remains is the insight Darwin argued already 1871 in his *The descent of man: 'No one, I presume, doubts that the large proportion which the size of man's brain bears to his body, compared to the same proportion in the gorilla or orang, is closely connected with his higher mental powers'* (Darwin, 1871). This assumption became the driving force for most of the work in the lab. The present book aimed to characterize mechanism to explain the expansion of the human neocortex as a first step towards a more composite multifactorial concept of how the neocortex developed its fascinating features and thus - about what makes us human.

7.2   In utero electroporation of a human-specific gene in the developing mouse neocortex

7.2.1 Opportunities and limitations of the approach

Due to ethical and legal restrictions, many procedures in experimental biology are not applicable to non-human primate or even human brains. Consequently, other tools or organisms have to be used to study the evolutionary benefits of a human-specific gene as it was the aim of this study. Fortunately, in utero electroporation (IUE) provides an established method to study a gene function at different developmental time points in different neocortical regions in vivo and offers the opportunity to examine the transient effects of a gene in a species, where it usually is absent. Albeit, the approach is based on the assumption that interactors and pathways of the protein are conserved between mouse and humans. If they are not, the functional role it has in humans remained uncovered, or an effect observed rather had to be assumed as an artifact than as the observation of a physiological phenomenon. The results we obtained in this study could be limited due to this issue since a knockdown study of mouse Fam72a showed an effect on neurogenesis, but overexpression of human FAM72A and FAM72D did not. IUE of a gene might increase the expression level of the transfected protein, but not of the entire biological machinery which is involved in the physiological function – an exception from this statement may be the IUE of a transcription factor. In contrast to these concerns are the reliable results of IUEs performed with ARHGAP11B showing a definite effect on the proliferation of basal progenitors (Florio et al., 2015), though it is not assumed to be a transcription factor. Nevertheless, it remains a general issue to be aware of when a molecular or cellular mechanism is studied in a species distantly related to the organism one aims to understand. Thus, the extrapolation from the mouse to the human brain can be problematic not only due to the remarkable differences regarding complexity and connectivity which are most likely not only caused by one single gene and consequently hard to reproduce in the existing experimental procedures. A related matter arises considering a more macroscopic perspective: several evolutionary adaptations might have mutually favored each other (see: 7.1.2) resulting in a minimized or even no effect when only one is induced in the model organism. Another critical point in the design of this study might be the use of the strong constitutive CAG promotor causing a gene dosage above the physiological level and not reflecting possibly important spatial and temporal dynamics, but an expression in neurons where it usually is absent.

A third limitation of the approach could be that non-mouse proteins provoke an immune reaction leading to the demolition of the transfected cells. Like any other protein of the cell, fragments of the artificially expressed unfamiliar human protein will be presented via the MHC I antigen presenting machinery at the surface of the cell causing an immune response via CD8+ cytotoxic T cells that induces apoptosis. The upregulation of Tapasin in mouse brain cells expressing FAM72D could be a hint towards this idea. Once a CD8+ T cell has recognized an alien protein fragment presented by MHC I molecules they start to secrete several cytokines including Inf-γ, which was found to upregulate Tapasin expression as a part of the MHC I antigen presenting machinery (see: Results IV), (Abarca-Heidemann et al., 2002; Tang et al., 2016). This process could also explain the upregulation of several genes involved in the NF-κB pathway as well as apoptosis. Most importantly it would cause a reduced number of FAM72A or FAM72D positive cells, and false negative results would be facilitated in this experimental setup.

Albeit, it has to be made clear that the use of IUE for experimental purposes like they were carried out in this study is widespread and established and might consequently only be accompanied by a few of the potential limitations discussed above.

7.3   The family of sequence similarity 72 and human uniqueness

In this book it was intended to characterize one candidate, a human-specific gene family in terms of its contribution to the expansion in neuronal number during neocortical development, which defines the human brain as the largest among primates (Herculano-Houzel, 2009). The following chapter will discuss potential effects of the FAM72 gene family on various parameters likely to affect neocortical growth.

7.3.1   Cell cycle regulation and NPC maintenance

The neuronal output can be increased when the pool of neural progenitor cells is enlarged before the switch to neurogenic cell divisions. Several studies have suggested that a lengthening of the cell cycle of NPCs in the VZ is accompanied by the progression of neurogenesis (Caviness et al., 1995; Takahashi et al., 1995). Following studies could further distinguish several NPC populations characterized by different cell cycle lengths and proportions of the sub-phases (Calegari et al., 2005; Gal et al., 2006; Arai et al., 2011; Florio & Huttner, 2014). Notably, the length of the G1 phase turned out to be a cell fate determinant, since a prolonged G1 phase in mice NPCs led to a premature neurogenesis at the cost of NPC pool expansion (Calegari & Huttner, 2003; Lange et al., 2009), whereas a shortening of G1 increased NPC proliferation accompanied by a delayed neurogenesis (Pilaz et al., 2009; Lange et al., 2009; Nonaka-Kinoshita et al., 2013). These insights are also evident in species evolutionary closer related to humans such as macaques, where G1 phase was identified to be shorter in cortical areas with cytoarchitectural hallmarks of neocortical expansion (Lukaszewicz et al., 2005; Florio & Huttner, 2014) supporting the idea, that G1 phase regulation is a significant factor in primate neocortex evolution. Interestingly, analyses of a FAM72A transfected nasopharyngeal carcinoma cell line suggest that FAM72A shortens G1 phase and increases the cell proliferation. Furthermore, it enhanced levels of cell cycle activators were found in western blot analyses upon FAM72A transfection (Wang et al., 2011). The finding that FAM72A effects the cell cycle was further supported by data collected in aphidicolin-synchronized FAM72A transfected MCF-7 cells showing that FAM72A shifts the transfected cells to a different cell cycle sub-phase compared to the control (Heese, 2013). Another study additionally attracted our attention to the FAM72 genes identifying Fam72a in adult mouse NPCs as one of the genes marked by the trimethylation of histone H3 at lysine 4 (H3K4me3) which is known to preferentially mark genes that are essential for the function and identity of the cell type (Benayoun et al., 2014). A lentiviral-based RNA interference knock-down of Fam72a in an adult primary mouse NPC cell culture led to increased neurogenesis suggesting that Fam72a might support the maintenance of NPC – a function which could result from the shortening of the G1 phase. Furthermore, NPC proliferation was slightly reduced upon Fam72a knock-down, though it is surprising that knock-down of Srgap2, a gene not yet described to be involved in the regulation of cell proliferation, affected NPC proliferation even stronger in this set of experiments (Benayoun et al., 2014), which either suggests a novel function of Srgap2 or is reason to question the validity of the results.

Albeit, as described in Results II and III neither expression of FAM72A nor FAM72D showed significant effects on the NPC proliferation or abundance of proliferative or neurogenic NPCs in the developing mouse neocortex. Further analyses (see: 7.4) will show whether the lack of a cell cycle-dependent phenotype in this set of experiments was rather due to the limitations of the approach or reflected the circumstance that the Family of sequence similarity 72 is not involved in the embryonic NPC cell fate regulation under physiological conditions in utero.

7.3.2   Cell death

Neocortical development is a precisely regulated process also characterized by a balance of proliferation and apoptosis. Interestingly, the highest rates of apoptosis were described at the peak of neurogenesis as well as in proliferative cortical zones (Blaschke et al., 1996; Thomaidou et al., 1997). Besides, the fundamental importance of apoptosis for healthy brain development was further dissected in experiments with caspase 3 or 9 deficient mice embryos, which showed an unphysiological expansion as well as exencephaly of the forebrain (Haydar et al., 1999). The existing data of FAM72A's involvement in the regulation of programmed cell death is conflicting to date, since on the one hand FAM72A was shown to be capable of inhibiting staurosporine-induced apoptosis in proliferating MCF-7 (breast cancer) cells (Heese, 2013), but on the other hand also identified to be potentially pro-apoptotic in rat neuroblastoma cells (Nehar et al., 2009). Our transcriptome data presented in Results IV shows several genes upregulated upon FAM72A or FAM72D which are involved in programmed cell death (e.g., *Stk17b, Shisa5, Parp 9, Irf1*). This could be either due to the suggested limitation of the approach resulting in the apoptosis-inducing immune reaction against the human proteins, but also a consequence of a physiological function of the two proteins. If it is true and FAM72 genes are involved in the sensitively regulated process of apoptosis during neocortical development, new experiments had to be executed considering the physiological temporal and spatial expression patterns of these proteins. These could also explain the controversy of the existing publications, possibly showing an effect which also depends on the cell type the FAM72 proteins are expressed in.

7.3.3 Neurogenic period

Another parameter determining the neuronal output is the neurogenic period. Indeed, interspecies variations have been reported between primates and rodents (Caviness et al., 1995; Rakic, 1995; Rakic, 2009; Florio & Huttner, 2014) and were shown to sufficiently explain differences in neocortical size and neuronal number between species of the same principal group (Lewitus et al., 2014).

One possible mechanism underlying a protracted neurogenic period could be an adapted temporal expression pattern of neurogenesis regulating proteins or the occurrence of extra copies of such regulators with a delayed expression pattern as it could be the case for the FAM72s. The findings discussed in 7.3.1. cell cycle regulation and NPC maintenance could support such a scenario. Besides, there are hints towards a sequential temporal expression pattern of the four paralogues, since amplification of *FAM72A, B, C, D* from a GW 12 human brain cDNA sample (see: 3.4.1) did only produce *FAM72A* copies, whereas *FAM72D* mRNA was detected wpc 13 (GW 15) (Florio et al., 2015). This temporal shift in expression is further supported by the Fietz dataset from 2012 which sampled four different gestational ages and showed that FAM72C and FAM72D are enriched in older samples (14-16 wpc) (Florio et al., 2018). This raises the possibility that they start to do the ancestral FAM72A job, when it is downregulated (further studies should dissect this hypothesis analyzing the temporal expression pattern of FAM72A) basically scaling up the chimpanzee brain (Herculano-Houzel, 2009) or could add a new function as discussed below (see: 7.3.4).

7.3.4 TrkB signaling

Over the past years the TrkB receptor was found to be a vital component of a signaling cascade affecting several fundamental aspects of brain development including the precursor cell survival, proliferation, neurogenesis and neuronal migration (Barnabe-Heider & Miller, 2003; Bartkowska et al., 2007; Puehringer et al., 2013). This is why the results of experiments performed with FAM72A transfected B104 (neuroblastoma) cells exhibiting an inhibited TrkB phosphorylation and TrkB downstream cascade upon BDNF stimulation (Nehar et al., 2009) attracted our attention. Although the finding itself should be reproduced in vivo in the developing neocortex at first, the analysis of our transcriptome data tempts to suggest the following model: Slitrk5 was among the genes upregulated explicitly upon FAM72D expression and previously found to interact with the TrkB receptor (see: Results IV). In fact, it is an important mediator of TrkB receptor recycling (Lazo et al., 2013; Song et al., 2015). This portends that the occurrence of FAM72D could have enabled the indirect rescue of the TrkB inhibition of FAM72A via Slitrk5 supporting the recycling of the receptor. Though, if FAM72A would inhibit TrkB in the developing mouse neocortex one could have expected a TrkB related phenotype upon the expression of either FAM72A or FAM72D in the analyses presented in Results II and III, it might still be worth to further dissect this coherence, since it cannot be excluded that the conducted experimental design was not connected to the fitting read-out revealing an existing phenotype. The negative results to date could be explained by the limitations of the approach described above, but also through various additional issues such as a different or even no interaction between human FAM72A and mouse TrkB, which would support the need of further studies in a different model organism evolutionary closer related to humans like ferret or chimpanzee derived brain organoids.

7.3.4.1 *Neuronal function*

Another aspect potentially connected to the interaction with the TrkB signaling pathway (Gupta et al., 2013) is the finding that several genes strongly associated to a neuronal function like synaptic plasticity such as Syde1 but also Slitrk5 were found to be differentially expressed upon FAM72A or FAM72D (Slitrk5) or both (Syde1). This was surprising considering that mRNA of all of the four paralogues was almost exclusively found to be expressed in NPCs but not neurons (Florio et al., 2018) suggesting a role in early neuronal differentiation. Thus, it provides another direction of future studies to examine the biological function of FAM72A-D in human neocortical development.

7.3.4.2 *Alzheimer's disease*

One more aspect to mention due to its potential intersections with various of the discussed points is an involvement in the pathophysiology of Alzheimer's disease. FAM72A was upregulated in an Alzheimer's disease mouse model (Nehar et al., 2009) and could be the missing link between the known effect of $A_{\beta 42}$ to decrease the activity of signaling pathways downstream of TrkB (Tong et al., 2004) mediating the subsequent interference with a normal BDNF function, possibly also in the adult hippocampus where the disease oftentimes begins and FAM72A was found to be expressed (Nehar et al., 2009).

7.3.5 Mitochondria

Several insights on the FAM72 gene family of previous but also this study suggest a link between FAM72s and mitochondrial functions, which will be discussed in the following.

7.3.5.1 Subcellular localization

As introduced in 1.6, FAM72A was found to be co-localized to mitochondria (Wang et al., 2011; Heese, 2013), a result which is supported by the biocomputational prediction of the FAM72s to possibly have a mitochondrial subcellular localization (http://www.cbs.dtu.dk/services/TargetP/ and http://mitf.cbrc.jp/MitoFates/cgi-bin/top.cgi). Unfortunately, we were not able to prove these results experimentally, since none of the tested FAM72 antibodies (see: 9.8) provided a specific signal in immunohistochemistry experiments under various fixation and staining conditions neither on in utero electroporated neocortical mouse nor on fetal human neocortex slices (data not shown). However, if FAM72A, B, C or D is co-localized to mitochondria, it would be particularly interesting to study whether there is a synergistic effect with ARHGAP11B, that was recently identified to be also co-localized to mitochondria (Namba et al., 2019).

7.3.5.2 Reactive oxygen species

It has also been Wang et al. (2011) who showed a decrease in reactive oxygen species (ROS) production induced by H_2O_2 treatment in FAM72A - compared to mock-transfected nasopharyngeal carcinoma cells. ROS are known to function as signaling molecules regulating a number of cellular processes including proliferation (Sundaresan et al., 1995; Behrend et al., 2003; Menon et al., 2007; Bartosz, 2009), cell cycle progression (Menon et al., 2007; Sarsour et al., 2009) and the self-renewal capacity of NPCs (Khacho & Slack, 2018). This is mentioned because of its possible relevance to explaining the effects of FAM72A on the cell cycle, but also due to the relation between ROS production and mitochondrial function.

7.3.5.3 Mitochondrial dynamics and neurogenesis

The general importance of mitochondria in the brain as the cellular energy producers is well established mainly due to the high-energy demanding nature of neurons, in contrast to the role of mitochondrial morphological dynamics, which was revealed in recent studies in post-mitotic neurons as well as in NPCs (Khacho & Slack, 2018). Mitochondria are highly dynamic organelles continuously cycling through fusion and fission events (Detmer & Chan, 2007) exhibiting distinctive morphologies in different cell types. The necessity of a well-orchestrated mitochondrial morphology is quintessentially seen in the association of abnormal mitochondrial dynamics and function to several neurodegenerative conditions (Oettinghaus et al., 2012) such as Alzheimer's, Parkinson's and Huntington's diseases (Wang et al., 2009; Calkins et al., 2011; Song et al., 2011; Wang et al., 2012), but also neurodevelopmental disorders causing impairments in brain development, since several studies could show an important role during NPC fate decisions (Steib et al., 2014; Khacho et al., 2016) and neuronal differentiation (Knobloch et al., 2013; Khacho et al., 2016; Khacho et al., 2017; Khacho & Slack, 2018). Interestingly, the mitochondrial morphology appears to be different in embryonic and adult

NPCs but is strongly related to relevant functions in both stages of development in the context of neurogenesis (Khacho & Slack, 2018).

7.3.5.4   *Mitochondrial fission protein 1*

As described in Results IV, the mitochondrial fission protein 1 (Mtfp1) was identified to be upregulated upon ectopic expression of the human-specific FAM72D but not the ancestral FAM72A. Mtfp1 was previously shown to be a key regulator of balanced fission and fusion of mammalian mitochondria, which puts it into the position of a potential regulator of human neocortical neurogenesis in concert with FAM72D. Albeit, the experimental design to further dissect this biological process might be challenging since the physiological phenotype is more likely to be seen as a result of a precisely regulated expression pattern of the proteins than in consequence of a simple gain or loss of function study. Whereas one way to get an orienting idea of Mtfp1's relevance for cortical neurogenesis surely could be a knock-down study in human-derived brain organoids particularly in the case of an additionally validated upregulation of Mtfp1 upon FAM72D expression or reduced level of expression upon FAM72D knock-down.

7.3.6   Angiogenesis

7.3.6.1   *Angiogenesis & neurogenesis*

Several studies could prove the interdependency of angiogenesis and neurogenesis from embryonic development until adulthood when whole body exercise-induced reduction of age-dependent cognitive decline can at least in part be traced back to a positive effect on angiogenesis (Stimpson et al., 2018).

Observations on the cellular level show that adult hippocampal neurogenesis occurs in a vascular niche, since proliferating cells are organized in dense clusters around small capillaries (Palmer et al., 2000; Ward & Lamanna, 2004), which mainly disappear after a couple of weeks suggesting that neurogenesis might be associated to an active vascular recruitment likely caused by the increased metabolic demand, but also by the need for "instructive cues" of the CNS-vascular interface (Palmer et al., 2000; Fuchs & Schwark, 2004; Yang et al., 2016). This view is supported by in vitro co-culturing experiments of embryonic neural stem cells with immortalized endothelial cells which increased the neural stem cell expansion and shifted their fate toward neurons (Shen et al., 2004). However, also data from in vivo studies confirm these results showing a remarkable temporo-spatial congruence of the induction of radial glia differentiation and intraparenchymal formation of vessels (Lange et al., 2016). This is consistent with results showing that the ventrolateral to dorsomedial gradient of periventricular vessel formation matches the gradient of neurogenesis and precedes the latter one by about one day (Vasudevan et al., 2008; Morante-Redolat & Fariñas, 2016).

7.3.6.2   Enriched GO term angiogenesis

Out of the 8 differentially expressed genes in the GO cluster angiogenesis, the most prominent candidate is a vascular endothelial growth factor, Vegf-c, which was already briefly characterized (see: Results IV).

Interestingly, Vegfs are generally known to exhibit direct effects on neurons including neurotrophic activity stimulating neurite outgrowth (Sondell et al., 1999), in addition to its angiogenic functions. Thus, in rats undergoing focal cerebral ischemia, administration of Vegf could enhance neurogenesis, cerebral angiogenesis and reduced the size of the infarct (Sun et al., 2003). Vegf-c, adds an exciting feature which is of high importance for several processes associated with immunological functions, such as lymphatic vessel development, for instance. But also, Thsd7a, the second gene in the GO cluster angiogenesis upregulated upon FAM72A and FAM72D expression, is linked to a variety of functions related to the endothelium. Considering the striking interaction between angiogenesis and neurogenesis, it is tempting to follow up this initial finding of the enriched GO term angiogenesis upon expression of FAM72A, which might also lead to results explaining why the GO enrichment was restricted to the ancestral paralogue of the FAM72 gene family. However, the natural temporal angiogenesis in mice may be a limitation to study effects upon FAM72A expression using IUE, since surface pial vessels already surround the entire brain at E 9.5 (Morante-Redolat & Fariñas, 2016) and vessel formation in the dorsolateral cortex occurs already between E 11.5 to E 12.5 (Lange et al., 2016). Hence, in utero electroporation to test FAM72s capability to induce or enhance this early angiogenesis at or before these time points, is technically extremely difficult due to the small size of the ventricle at this developmental stage and might cause the need to use a different model system.

7.3.7 An evolutionary immunological adaptation in the brain?

Over the past years, tremendous progress was made in the understanding of the interaction between two of the most complex and dynamic systems in our body – the immune and the nervous system. The results from our transcriptome analysis identified several genes differentially expressed upon FAM72A or FAM72D known to be involved in immunological and inflammatory often NF-κB related pathways including *Adora2b, Shisa5, Stk17b, Syde1, Tnfrsf13c, Ptx3, Kctd6* (see: Results IV).

Interestingly, 3 out of 6 specifically upregulated genes upon FAM72D expression are experimentally found to be associated with some immunological response: *Parp9, Cnp* and *Tapasin*.

Heterodimerization of Parp9 with Dtx3L improves the antiviral response, likely due to indirect stimulation of interferon-stimulated genes expression (Zhang et al., 2015). A lack of Cnp was found to increase levels of inflammatory mediators and higher levels of Cnp were detected in activated microglia cells (Yang et al., 2014). The third Tapasin is involved in one of the most fundamental immunological processes: the antigen processing and presentation machinery of the cell. Although its upregulation could be the result of the proposed immune reaction against the human protein in the mouse organism (see: 7.2), it would be hastily to consider the finding as such an artifact. One intersection could exist between Tapasin and MICA (MHC class I polypeptide-related sequence A), a gene recently identified to be primate specific and to be expressed in cortical NPCs (Florio et al., 2018). It is a cell surface glycoprotein located within the MHC locus and is related to MHC class I molecules exhibiting a similar domain structure (Bahram et al., 1994), but is not associated to the $ß_2$-microglobulin. This suggests together with the finding of normal MICA surface levels in cells lacking the TAP peptide loading transporter, the supplier of peptides bound by class I molecules from the cytosol into the lumen of the endoplasmic reticulum, a functional independence of MICA of cytosolic peptide ligands (Groh et al., 1996), in contrast to MHC class I molecules. Hence, Tapasin and MICA may not directly interact but could be part of a more complex immunological adaptation that occurred throughout primate evolution in the developing brain.

7.3.8   FAM72 and SRGAP2

The genomic neighborhood of the *FAM72s* and the *SRGAP2s* due to the paired duplication during human evolution provided the opportunity to estimate the duplication time points of the *FAM72* paralogues based on the data from the *SRGAP2* paralogues between 3.4 – 1 mya, (Dennis et al., 2012; Kutzner et al., 2015) – a period of time corresponding to the emergence of the genus Homo, the expansion of the neocortex as well as the use of stone tools and other striking cultural changes (Jobling et al., 2004). Previous work on SRGAP2A could show that it negatively regulates neuronal migration and induces neurite outgrowth (Guerrier et al., 2009). Subsequent work including the human-specific paralogues found that SRGAP2C exhibits an inhibitory effect, due to a truncated F-BAR domain, on the ancestral SRGAP2 function consequently accelerating neuronal migration. Additional experiments showed a spine morphology shaping activity of SRGAP2 resulting in a delayed spine maturation and an increased spine density upon SRGAP2C expression (Charrier et al., 2012) possibly underlying this and further substantial differences observed between human's and other species' spine morphologies. Those have been proposed to introduce more flexibility for input integration and information processing with impact on human cognition, learning and memory (Benavides-Piccione et al., 2002; Charrier et al., 2012). These and other intriguing findings like an involvement in the regulation of excitation and inhibition in the brain (Fossati et al., 2016; Subramanian & Nedivi, 2016) and genetically caused brain abnormalities upon deletion (Rincic et al., 2016) brought up the idea of the *brain-specific neural master gene pair SRGAP2-FAM72* (Ho et al., 2017a). The authors of this hypothesis propose that NPCs activate *FAM72s* to proliferate and switch to SRGAP2 expression when the cell decides to differentiate into neurons and starts with migration and synaptogenesis (Ho et al., 2017a; Ho et al., 2017b). Future studies could elucidate whether FAM72s and SRGAP2s indeed are sequentially expressed.

The published data to date indeed suggests the concept of the *FAM72-SRGAP2* master gene, which underlies our higher cognitive functions. Our data, however, prompts to be more careful with such statements, at least in the case of the *FAM72* gene family since the underlying biological mechanisms seem to be more complicated. Our results and a lack of sufficient evidence in previous studies (Nehar et al., 2009; Heese, 2013; Benayoun et al., 2014) rather question the claimed (Ho et al., 2017b) pivotal role of the FAM72A or FAM72D protein for embryonic NPC proliferation, without excluding it, but asking for additional experiments to bring the tempting, though still very primitive hypothesis, that the occurrence of the *FAM72-SRGAP2 master gene* could underlie the evolution of human cognition on safer ground. An issue supporting the evolutionary relevance of the FAM72s is that in contrast to the SRGAP2s (Dennis et al., 2012) none of the paralogues became a pseudogene (Florio et al., 2018) suggesting a beneficial evolutionary function.

7.3.9 FAM72, Neanderthals, and lncRNAs

Consistent with the estimated duplication time points FAM72A, B, C and D are also present in the genome of Neanderthals and Denisovans. Sequence comparison of the Homo sapiens to the Neanderthals and Denisovans genome revealed identical exons but displayed variations in the introns of human compared to Neanderthal and Denisovan *FAM72A* and FAM72*D* (Kutzner et al., 2015). Analyses of the genomic locus and more specifically, the interspace between the *FAM72* and *SRGAP2* genes led to the discovery of homologous long non-coding RNAs (lnc RNAs) separating the four gene pairs of *SRGAP2* and *FAM72*. This opens an exciting new field not considered in the experiments of this study, since all data is based on the cDNA sequence lacking the introns of *FAM72A* and FAM72*D* as well as the sequences from the *FAM72 – SRGAP2* interspace, due to the focus on the characterization of a putative protein function during neocortical development. Recent insights in the variety of functional roles lncRNAs play during neocortical development (Hart & Goff, 2016; Florio et al., 2017; Heide et al., 2017) support future experiments to elucidate the importance of these non-protein-coding sequences within an evolutionary highly interesting region of the human genome.

7.4 Future directions

Although a great variety of different paths for future studies could be created, the following suggestions will be restricted to a selection of experiments designed to elucidate the most critical questions that occurred upon the results of this study and those likely to be performed due to the existing know-how and technical opportunities in the lab.

7.4.1 Loss of function

Our main conclusion from Results II & III has been, that FAM72A and FAM72D are not sufficient, but may be required to enhance NPC proliferation in the developing mouse neocortex. Hence, a loss of function study is an important next step to do.

This set of experiments could be performed in utero to reproduce the findings of Benayoun et al. (2014) showing increased neurogenesis but decreased NPC proliferation upon shRNA mediated knock-down of *Fam72a* in adult NPCs. Conducting a similar experiment during embryonic development could give a first interesting hint if the functional role of Fam72a is different between the embryonic and adult stage in NPCs. Following rescue IUE with human FAM72A or FAM72D or mouse Fam72a could identify a neofunctionalization throughout evolution in the case that IUE of Fam72a rescues the phenotype in contrast to IUE of FAM72A or FAM72D.

These results could be further dissected using human-derived brain organoids where the *FAM72* genes are silenced or even knocked out via CrisprCas9, for instance. One particular challenge of the latter approach might be to selectively downregulate only one of the paralogues, although it would still be interesting to analyze the effects the lack or downregulation of the entire gene family exhibits. Despite the significant progress in the organoid field during the past years, a second limitation might be the variability and quality of the organoids, which can be grown to date (Di Lullo & Kriegstein, 2017). As a first step, it may be most reasonable to design the readout similar to the presented Results II & III.

7.4.2 Gain of function

Intuitively, it also makes a lot of sense to study the effects of genes that newly arose during evolution in the setting of a gain of function approach, as it was performed in this study. Albeit, as discussed above (see: 7.2) this comes with a row of issues related to the relatively long evolutionary distance between mouse and humans. Thus, it would be a good experiment to reproduce a specific part of human evolution as close as possible to what happened during the last 4 to 1 million years via the expression of the human-specific paralogues in chimpanzee derived brain organoids.

In this study, I only focused on the functional characterization of *FAM72A* and *FAM72D*. IUE of the human-specific paralogues *FAM72B* and *FAM72C* in mice embryos or chimpanzee derived brain organoids could unveil a neofunctionalization of these two paralogues. Furthermore, it would be interesting to screen for the short isoforms of these paralogues found to be expressed in the developing neocortex (Florio et al., 2018). Due to the enormous amount of work to quantify this set of experiments, we tried to establish a flow cytometry protocol for automated quantification of the different markers but did not succeed at the time. Albeit, it could be worth to reinvest some efforts in this technique because it would provide not only a faster but also more objective alternative to manual counting. Another way to improve the quantification process could be the use of software which might soon be available considering the rapid progress in the field of bioinformatics to reliably quantify cell numbers from immunofluorescence images.

Finally, the possibility that the interaction of some of the human-specific genes is required to produce the phenotype observed in the brain of modern humans should be considered. Hence, co-electroporation of different theoretically promising combinations of the candidate genes in chimpanzee brain organoids as well as mouse embryos would be an exciting experiment to do. One example for such an experiment could be the co-electroporation of *ARHGAP11B* and *FAM72D* based on the rationale that both are assumed to be co-localized to mitochondria (Wang et al., 2011; Namba, 2019) and might functionally interact or to co-electroporate *FAM72A, B, C* and *D* together to study an interaction like it is observed in the case of SRGAP2A and C (Charrier et al., 2012). Based on the premise that the human brain can basically be understood as a scaled-up primate brain (Herculano-Houzel, 2009), it is reasonable to assume that a significant part of the human neocortical expansion is the result of an adaptation of genes already existing in the primate genome. Thus, a logical conclusion would be to include those genes expressed in human fetal NPCs identified to be primate-specific to the analyses (Florio et al., 2018).

Moreover, a previously identified till now only little considered aspect is worth to be followed up to be taken into consideration: the interaction with the TrkB signaling pathway (Nehar et al., 2009). The major questions to solve here are, if FAM72A inhibits TrkB receptor signaling also in vivo in the developing neocortex and if yes, it has to be identified, whether FAM72D exhibits similar inhibitory effects. In parallel, and particularly in the case that FAM72D does not inhibit TrkB signaling, a validation of the Slitrk5 upregulation upon FAM72D expression may be helpful as a starting point to dissect whether FAM72D indeed could indirectly rescue the inhibition of FAM72A via the upregulation of Slitrk5.

Finally, an aspect to check is the potential interaction with mitochondrial dynamics of FAM72D via Mtfp1. As in the case of Slitrk5, a first step would be to validate the differential expression found in our transcriptome analysis using a qPCR, for instance. If this is the case, it opens an exciting field for further studies which aims to understand how mitochondrial morphology changed over the course of primate evolution in NPCs and how it differs during embryonic development of an individual.

8   Summary / Zusammenfassung

8.1   Summary

Introduction: The higher cognitive functions that characterize modern humans can be attributed to the cerebral neocortex and its remarkable expansion in size during the last 5 – 7 million years of human evolution. The identification of the underlying genomic changes will be not only of importance to better understand the unique complexity of the human brain, but also its susceptibility to neurological and psychiatric diseases.

Recently, 15 human-specific genes preferentially expressed in neural progenitor cells (NPCs) of the human fetal neocortex were identified (Florio et al., 2018). Three of them, *FAM72B, C* and *D* belong to the *Family of sequence similarity 72 (FAM72)* and occurred in the human genome by gene duplication 3.4 – 1 mya.

Aims & Approaches: Specifically, it was asked whether FAM72D plays a diverse role compared to the ancestral FAM72A (Results II, III, IV) due to the specific sets of amino acid substitutions it acquired (Results I). Effects of FAM72A and FAM72D on the proliferative capacity and gene expressions of embryonic mouse NPCs were analyzed upon ectopic expression either of FAM72A or FAM72D during embryonic mouse neocortical development.

Methods: In utero electroporation (IUE) of embryonic mouse brains was performed to drive the expression of a red or green fluorescent protein (RFP or GFP) either plus empty DNA vector (pCAGGS; control), pCAGGS-*FAM72A* or pCAGGS-*FAM72D* plasmids in the dorsolateral neocortex at mid-neurogenesis (embryonic day 13.5, E13.5; Results II) or in the medial neocortex at late-neurogenesis (E15.5; Results III). NPC proliferation was evaluated by immunofluorescence of Ki67 (immunohistochemistry, IHC), a cell proliferation marker, and phosphorylated Histone H3 (PH3), a marker of cell mitosis. Moreover, the abundance of NPCs using immunofluorescence of basal intermediate progenitor (Tbr2) and apical and basal radial glia (Sox2) markers, and the gliogenesis by Olig2 immunofluorescence was analyzed. Additional experiments were carried out to study the capacity of NPCs to reenter the cell cycle upon IUE of FAM72D. To this end, pregnant mice were intraperitoneally injected with the thymidine analog 5-Ethynyl-2´-deoxyuridine (EdU) 24 h post-IUE, to label all cells undergoing S-phase of the cell cycle (i.e., all cells that reentered the cell cycle after IUE) in the developing mouse brains. Embryonic brains were collected 24 h after EdU injection and co-stained with Ki67. Ki67 and EdU double positive cells were considered as cells that reentered the cell cycle. To execute the transcriptome analysis E13.5 mice were electroporated with pCAGGS-*GFP* either plus an empty DNA vector (pCAGGS, control), a vector driving expression of FAM72A (pCAGGS-*FAM72A*) or FAM72D (pCAGGS-*FAM72D*). Subsequently, the electroporated dorsolateral neocortical areas were microdissected at E14.5 and dissociated into single cells. The electroporated (GFP+) cells were isolated from the single cell suspensions by the fluorescence-activated cell sorting (FACS). The isolated cells were processed for RNA sequencing. Data analysis was performed as previously reported (Florio et al., 2015).

Results: By immunohistochemistry, no significant changes in any of the proliferative parameters or in the abundance of progenitors in the ventricular zone (VZ) and subventricular zone (SVZ) of the developing mouse neocortex upon ectopic expression of FAM72D compared to FAM72A and control samples were detected (Results II, III).
However, the transcriptome analysis (Results IV) showed 88 significantly up- and 52 down-regulated genes upon FAM72A and 91 significantly up- and 67 downregulated genes upon FAM72D expression compared to the control. Only two of these differentially expressed genes were found to be upregulated upon FAM72A and FAM72D with an expression >1 fpkm: *Syde1* and *Shisa5*. Besides, six genes specifically upregulated upon ectopic expression of FAM72D exhibiting fpkm > 1 were identified and characterized using the existing literature: *Tapbp, Mtfp1, Slitrk5, Parp9, Cnp, Rbm43*. Beyond that, gene ontology analysis showed significant enrichment of angiogenesis-related genes (e.g., Vegfc) upon FAM72A expression. Interestingly, there were more genes found to be enriched in NPCs that were upregulated compared to control upon FAM72D than FAM72A expression, but more NPC enriched genes downregulated upon FAM72A compared to FAM72D expression. In the case of differentially expressed neuron-enriched genes, the data was were inverse, which slightly supports the idea that FAM72D rather than FAM72A could positively affect the maintenance of NPC characteristics.

Conclusions: In a previous study knockdown of Fam72a in adult mouse NPCs increased neurogenesis (Benayoun et al., 2014). This suggests, in conjunction with the present results, that FAM72A and FAM72D are not sufficient, but may be required, to promote NPC maintenance (Results II, III). This is why the gain of function experiments conducted in this study should be complemented by a loss of function approach in the developing mouse neocortex, in chimpanzee or human-derived brain organoids. Because of their expression in the NPCs of the developing human neocortex, it might be productive to analyze the potential synergistic effect on NPC proliferation of the *FAM72s* and the 12 other human-specific genes such as *ARHGAP11B*. Among other mechanisms discussed based on the gene expression analysis in this book (Results IV and Discussion), the upregulation of Slitrk5 upon ectopic expression of the human-specific *FAM72D* could be particularly remarkable. Slitrk5 is known to be involved in the recycling of the TrKB receptor (Song et al., 2015), which affects fundamental aspects of brain development. While FAM72A was found to inhibit the TrKB receptor (Nehar et al., 2009), the occurrence of FAM72D could indirectly rescue the TrKB receptor function via Slitrk5 and thereby prolonging or enhancing essential features such as precursor cell survival and neurogenesis in humans.

Therefore, this study provides the first functional characterization of the evolutionary highly interesting region in the human genome comprising the *FAM72* genes during embryonic neocortical development in vivo and offers numerous starting points for further investigations, that will collectively facilitate a comprehensive understanding of the genomic adaptations underlying the astonishing evolution of the human brain.

8.2 Zusammenfassung

Einführung: Eine entscheidende Ursache für das Aufkommen der den modernen Menschen charakterisierenden kognitiven Funktionen ist in der beachtlichen Vergrößerung des menschlichen Neocortex innerhalb der letzten 5-7 Millionen Jahre zu finden. Die Identifizierung der dieser Entwicklung zu Grunde liegenden genomischen Veränderungen ist letztlich nicht nur bedeutsam für die Beantwortung der Frage, welche evolutionären Anpassungen den Menschen kennzeichnen, sondern auch für ein besseres Verständnis einer möglicherweise besonderen Anfälligkeit gegenüber neurologischen und psychiatrischen Erkrankungen.

Kürzlich konnten 15 menschenspezifische Gene, deren Expression sich vorzugsweise in neuronalen Vorläuferzellen (NPCs) des menschlichen fetalen Neokortexes nachweisen lässt, identifiziert werden (Florio et al., 2018). Drei davon (*FAM72B, C* und *D*) sind vor 3,4 – 1 Millionen Jahren im menschlichen Genom durch Genduplikationen entstanden und gehören zur *Family of sequence similarity 72 (FAM72)*.

Zielsetzung und Ansätze: Konkret wurde betrachtet, ob *FAM72D* durch die spezifischen Substitutionen von Aminosäuren eine sich von der Funktion des anzestralen Gens *FAM72A* unterscheidende Rolle in der neokortikalen Entwicklung einnimmt. Untersucht wurden deshalb die Effekte von FAM72A und D auf die Proliferationskapazität und Genexpression von NPCs nach der ektopen Expression von FAM72A oder D während der embryonalen Entwicklung des Neocortex der Maus.

Methoden: Die in utero Elektroporation (IUE) embryonaler Mäusegehirne erfolgte zur Expression eines rot oder grün fluoreszierenden Proteins (RFP oder GFP) entweder gemeinsam mit einem leeren DNA pCAGGS Vektor als Kontrollbedingung oder aber einem pCAGGS-*FAM72A* oder pCAGGS-*FAM72D* Plasmid. Die in der zweiten Ergebnissektion (Results II) präsentierten IUE wurden dabei im dorsolateralen Neokortex zum Höhepunkt der Neurogenese am 14. Entwicklungstag (E 14.5) durchgeführt, im Unterschied zu den Experimenten in der dritten Sektion (Results III), die im medialen Neokortex am 18. Entwicklungstag (E 18.5) während der Spätphase der embryonalen Neurogenese realisiert wurden. Die Proliferation der NPCs wurde durch Immunfluoreszenzanalysen zweier Marker (Ki67 und phosphoryliertes Histon 3) bestimmt. Zudem wurde die Häufigkeit wichtiger Subtypen von NPCs ebenfalls durch Immunfluoreszenzanalysen eines Markers für basale intermediäre Vorläuferzellen (bIPs → Tbr2) sowie für basale und apikale radiale Gliazellen (aRGs, bRGs → Sox2) ermittelt. Die Gliogenese wurde durch Olig2 Immunfluoreszenz quantifiziert. Weitere Experimente wurden durchgeführt, um die Fähigkeit der NPCs, den Zellzyklus nach der IUE von FAM72D erneut einzuleiten, zu untersuchen. Zu diesem Zweck wurde schwangeren Mäusen 24 h nach der IUE das Thymidin-Analogon 5-Ethinyl-2'-desoxyuridin (EdU) intraperitoneal injiziert. Damit wurden alle Zellen markiert, die sich zu diesem Zeitpunkt in der S-Phase des Zellzyklus befanden und damit den Zellzyklus nach der IUE fortsetzten. Nach weiteren 24 h (48 h post-IUE) erfolgte die Auswertung: alle Ki67- und EdU-doppelt positiven Zellen wurden als solche betrachtet, die den Zellzyklus nach IUE fortführten (EdU+) und nach weiteren 24 h noch immer proliferierten (Ki67+).

Zur Durchführung der Transkriptomanalyse wurden Mäuse am 13. Entwicklungstag mit pCAGGS-GFP und entweder dem leeren DNA-Vektor (pCAGGS, Kontrolle) oder einem die Expression von FAM72A (pCAGGS-FAM72A) oder FAM72D (pCAGGS-FAM72D) ermöglichenden Vektor elektroporiert. Anschließend wurden die elektroporierten dorsolateralen neokortikalen Bereiche am 14. Entwicklungstag mikroskopisch seziert und in einzelne Zellen dissoziiert. Die Isolation der elektroporierten (GFP+) Zellen erfolgte aus den Einzelzellsuspensionen durch Fluoreszenz-aktivierte Zellsortierung (FACS). Im Anschluss wurden die isolierten Zellen für die RNA-Sequenzierung vorbereitet. Die primäre Datenanalyse der Ergebnisse der RNA-Sequenzierung wurde entsprechend etablierter Protokolle durchgeführt (Florio et al., 2015).

Ergebnisse: Die Analyse der Immunfluoreszenzquanitfizierungen (Results II und III) ergab keine signifikanten Veränderungen der proliferativen Parameter oder der Häufigkeit der NPCs in der ventrikulären Zone (VZ) oder subventrikulären Zone (SVZ) des sich entwickelnden Mausneokortex nach der ektopen Expression von FAM72A oder FAM72D im Vergleich zur Kontrollbedingung.

Die Transkriptomanalyse (Results IV) zeigte jedoch 88 signifikant hoch- und 52 herunterregulierte Gene in Folge der FAM72A sowie 91 signifikant hoch- und 67 herunterregulierte Gene nach der FAM72D Expression im Vergleich zur Kontrolle. Es wurde festgestellt, dass nur zwei dieser differentiell exprimierten Gene in Folge der ektopen Expression sowohl von FAM72A als auch FAM72D hochreguliert wurden und ein Expressionslevel > 1 fpkm aufwiesen: *Syde1* und *Shisa5*. Darüber hinaus wurden sechs Gene mit > 1 fpkm identifiziert, die spezifisch nach der Expression von FAM72D hochreguliert waren: *Tapbp, Mtfp1, Slitrk5, Parp9, Cnp, Rbm43*. Darüber hinaus zeigte die Genontologie-Analyse (Gen Ontology) eine signifikante Anreicherung von Angiogenese-assoziierten Genen (z. B. *Vegfc*) im Datensatz der artifiziell FAM72A exprimierenden Zellen. Interessanterweise konnte beobachtet werden, dass unter den im Vergleich zur Kontrolle differentiell exprimierten Genen mehr Gene mit typischer Expression in NPCs in Folge von FAM72D als FAM72A Expression hochreguliert und mehr NPC typische Gene nach FAM72A Expression herunterreguliert wurden. Im Falle der Gene, deren Expression eher in Neuronen zu finden ist, zeigte sich ein entgegengesetztes Bild (Results IV). Diese Befunde lassen den vorsichtigen Schluss zu, dass

FAM72D stärker als FAM72A die Aufrechterhaltung von NPC-Eigenschaften positiv beeinflussen kann.

Schlussfolgerungen: In einer früheren Studie erhöhte der Knockdown von Fam72a in NPCs erwachsener Mäuse die Neurogenese (Benayoun et al., 2014). Dies legt in Verbindung mit den vorliegenden Ergebnissen nahe, dass FAM72A und FAM72D nicht hinreichend, möglicherweise jedoch notwendig sind, um die Aufrechterhaltung des Vorläuferzellcharakters von NPCs zu fördern (Results II, III). Aus diesem Grund sollte das in dieser Studie verfolgte *Gain of Function* Design durch einen *Loss of Function* Ansatz ergänzt werden. Als Modellsystem bieten sich hierfür der embryonale Mausneokortex oder Hirnorganoide aus Stammzellen des Schimpansen oder Menschen an.

Da alle der kürzlich identifizierten menschenspezifischen Gene in den gleichen NPCs exprimiert werden, sollte auch die potenzielle synergistische Wirkung auf die NPC-Proliferation der FAM72 und der zwölf anderen humanspezifischen Gene wie etwa ARHGAP11B analysiert werden. Neben anderen möglichen Mechanismen, die auf Grundlage der Genexpressionsanalyse im Diskussionsteil dieser Arbeit (Results IV und Discussion) erörtert wurden, könnte die Hochregulierung von *Slitrk5* in Folge der ektopen Expression des humanspezifischen FAM72D besonders relevant sein. Es ist bekannt, dass Slitrk5 am Recycling des TrKB-Rezeptors beteiligt ist (Song et al., 2015), der wiederum grundlegende Aspekte der Gehirnentwicklung beeinflusst. Ebenfalls konnte bereits gezeigt werden, dass FAM72A die Funktion des TrKB-Rezeptors hemmt (Nehar et al., 2009). Somit ist denkbar, dass FAM72D im menschlichen Neokortex die Wiederherstellung der TrKB-Rezeptorfunktion indirekt über Slitrk5 verbessert und dadurch wesentliche Parameter wie das Überleben von Vorläuferzellen und die Neurogenese beim Menschen verlängern oder verstärken könnte.

Diese Studie stellt damit die erste funktionelle Charakterisierung der evolutionär hochinteressanten, die FAM72 Gene beinhaltende Region des menschlichen Genoms während der Entwicklung in utero dar. Daraus ergeben sich zahlreiche Ansatzpunkte für zukünftige Untersuchungen, die in ihrer Gesamtheit ein umfassendes Verständnis der Evolution des menschlichen Gehirns ermöglichen werden.

9   Materials and Methods

9.1   Chart of all experiments

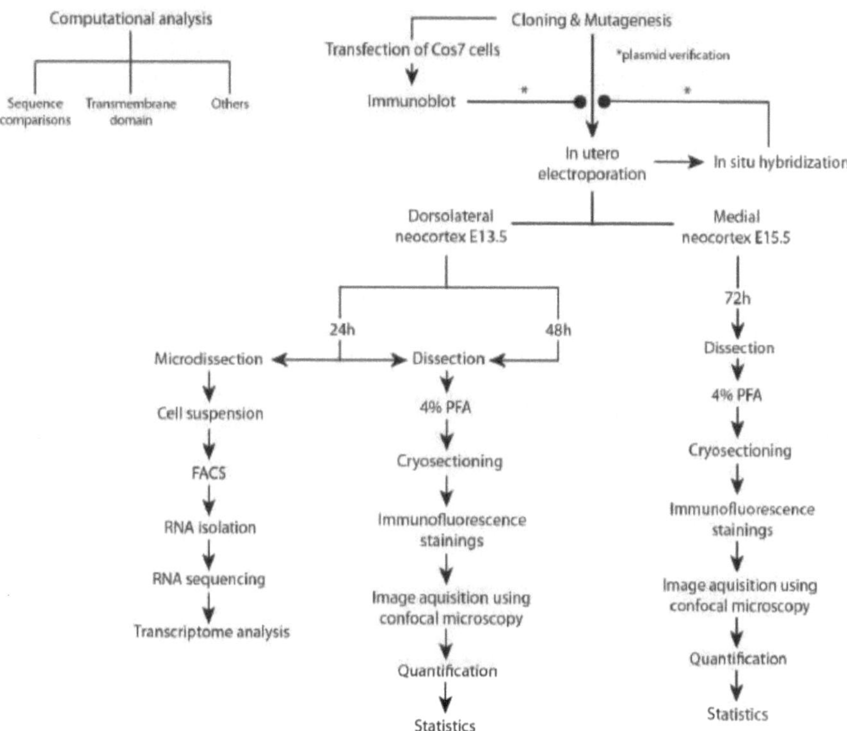

Fig. 29 – Chart of all experiments

This figure provides an overview of the experimental workflows in this study.

9.2   Computational analysis

9.2.1   Reference sequences and multiple sequence alignments

The following reference sequences were acquired from ENSEMBL.org and used for the initial sequence analysis as well as the plasmid design.

Table 16: FAM72 reference sequence ENSEMBL IDs

GENE NAME	ENSEMBL ID
FAM72A	ENSG00000196550
FAM72B	ENSG00000188610
FAM72C	ENSG00000263513
FAM72D	ENSG00000215784

The multiple sequence alignments were conducted using CLUSTALW http://www.genome.jp/tools-bin/clustalw.

9.2.2 Transmembrane domain prediction

Two different prediction tools were used to determine the existence of a transmembrane domain in the FAM72 proteins: firstly, the same one already used in previous analyses (Heese, 2013): http://harrier.nagahama-i-bio.ac.jp/sosui/sosui_submit.html and secondly, from the center for biological sequence analysis: http://www.cbs.dtu.dk/services/TMHMM/.

9.3   Amplification, subcloning, mutagenesis

9.3.1 Amplification from human brain cDNA

The FAM72 DNA was produced via amplification from a GW12 human brain cDNA samples as a template. The reaction mix consisted of 35 µl of double distilled water, 5 µl 10x KOD buffer, 5 µl dNTPs, 2 µl MgSO$_4$, 1 µl template, 1 µl primer (10 pmol / µl) and 1 µl KODplus (Toyobo). In total, I used three different primers: two different forward primers to amplify FAM72A/B and FAM72C/D, respectively and one reverse primer (Table 17). After 2 min of initialization at 94 °C, 40 cycles PCR (denaturation 94 °C: 15 s, annealing 65 °C: 30 s, elongation 68 °C: 1min) were performed and finished with a final elongation for 3 min, before the PCR product was kept at 4 °C.

Table 17: Primers used to amplify FAM72A, B, C and D

The red c in the FAM72 – 1F C/D sequence highlights the difference between the two forward primers.

Oligoname	Sequence	Amplification of
FAM72 – 1Forward A/B	ggGGATCCGCCACCatgtctaccaacatttgtag	FAM72A and B
FAM72 – 1Forward C/D	ggGGATCCGCCACCatgtccaccaacatttgtag	FAM72C and D
FAM72 – 1Reverse	cccGGATCCttatctaatacactcctctg	FAM72A, B, C and D

9.3.2 Subcloning

The amplified FAM72 DNA sequences were subsequently subcloned into TOPO vectors. Therefore, 2 µl of the PCR product, 0.5 µl salt solution, 0.5 µl pCR – BluntII- TOPO (addgene) were mixed and kept at room temperature for 5 min, before 50 µl of DH5a competent cells were added and the entire mix put on ice for 30 min. To transfect the DH5a cells, the mix was held for 35 s in a 42 °C water bath and directly put back on ice afterward for 1-2 min. Later, 150 µl LB medium was added and the entire mix incubated at 37 °C for 1 h. Finally, the solution was plated on a kanamycin agar plate to select those DH5a colonies which were transfected with the TOPO-*FAM72* plasmids.

After one day of incubation, several colonies were selected, diluted in LB medium that contains the antibiotic drug TOPO provides a resistance gene for (ampicillin) and incubated at 37 °C for one more day to increase the amount of *FAM72* DNA. The TOPO-*FAM72A/B/C/D* DNA was purified using the QIAGEN Miniprep Kit according to the producer's instructions. The final product was Sanger sequenced to validate the FAM72*A/B/C/D* DNA sequence.

9.3.3 Mutagenesis

Because only FAM72A copies were amplified from the human brain cDNA samples, I had to produce the three human-specific paralogues using mutagenesis. Therefore, the Mutagenesis-QuikChange Lightning Multi Site-Directed Mutagenesis kit (Agilent) was used according to the manufacturer's instructions. The primers used to induce the mutations are listed in Table 18. For additional illustration check Fig. 9. To perform mutagenesis on *FAM72A* and *FAM72A6ntC* (C instead of T at nt position 6) to produce *FAM72B*, *C* and *D* both *FAM72A* versions were subcloned into a TOPO vector. Afterward, starting with TOPO-*FAM72A*, the two substitutions of *FAM72B* were inserted by mutagenesis. Finally, TOPO-*FAM72C* was cloned from TOPO-*FAM72D* firstly obtained from the TOPO-*FAM72A6ntC* human brain cDNA PCR product (Fig. 9). After the sequences of TOPO-FAM72*A*, *B*, *C* and *D* were verified again by Sanger sequencing each of them was subcloned into a pCAGGs-vector which was used for all further experiments. The pCAGGs vector contains a CAG-promotor which constitutively strongly drives gene expression in mammalian expression vectors.

Table 18: Mutagenesis primers

The nucleotides inserted to perform the mutagenesis are written in capital letters.

Mutation	Primer sequence
FAM72B-281	gttcctgtcttcCttcctgcaacaa
FAM72B-364	caggtgtaaacAtcctactttgggg
FAM72D-295	cctgcaacaacAgacacttctgga
FAM72D-373	cgtcctacttCggggcaacttgcca
FAM72C-245	gtgggaacattgtagTttatcatgtg

9.4 Plasmid verification

9.4.1 Transfection of Cos7 cells

To transfect Cos7 cells with empty pCAGGS or pCAGGS-FAM72A/B/C/D the Amaxa device (Lonza) with program "W-001" was used. After trypsinization the cells were spinned down and collected. In In following 99 µl of NF solution (NucleofectorTM Solution, Lonza) and 1 µl of the plasmids (1 µg/µl) were added and the entire mix gently suspended using a pipette with small tip. Then the mixture was transferred to the Cuvette and 900 µl of DMEM (Dulbecco's Modified Eagle Medium, ThermoFisher) + FBS (Fetal Bovine Serum, Sigma-Aldrich) added. After electroporation (see: 9.6), cells were transferred to the 6 well plate (total 1ml per well). The transfection was conducted in collaboration with Takashi Namba (Huttner Group, MPI CBG).

9.4.2 Immunoblots

Whole cell lysates were prepared from transfected Cos7 cells by 1x SDS sample buffer and separated by SDS-PAGE (Nupage 4-13% gel, ThermoFisher) and transferred to membranes (XCell SureLock™ Mini-Cell Electrophoresis System, ThermoFisher).
After transfer, membranes were incubated in blocking buffer (5% BSA, bovine serum albumin in PBS) overnight at 4 °C. Incubation with primary antibodies against FAM7A/B/C/D (rabbit anti-FAM72A abbexa; abx 145553; 1: 5000) was performed for 1 h at room temperature. As a loading control, rabbit antibody against β-actin (1:5000, 4970, CST) was used. All antibodies were diluted in blocking buffer (5% BSA/PBS). Antigen-antibody complexes were detected using appropriate HRP-conjugated secondary antibodies (1:5000) and visualized with Super

Signal WestDura reagents (ThermoFisher). Membranes were exposed to Hyperfilm ECL (Amersham). The IB was conducted in collaboration with Takashi Namba (Huttner Group, MPI CBG).

9.4.3 In situ hybridization

In situ hybridization (ISH) was performed on 14 µm cryosections of E15.5 mouse brain slides in utero electroporated with pCAGGS-FAM72D expression plasmids. Before the hybridization step, cryosections were sequentially treated with 0.2 M HCl (2 × 5 min, room temperature) and then with proteinase K (6 µg/ml) in PBS, pH 7.4 (20 min, room temperature). Hybridization was performed overnight at 65 °C. TSA Plus DIG detection Kit (Perkin Elmer) was used for signal amplification, and the signal was detected immunohistochemically with mouse anti-digoxigenin HRP antibody (Perkin Elmer) and NBT/BCIP (Roche) as color substrate. For a more detailed description see Florio et al. (2018). The ISH was conducted in collaboration with Michael Heide (Huttner Group, MPI CBG).

9.5 Mice

All animals used in this study were wild-type (wt) C57BL/6JolaHsd mice. Noon of the day when the vaginal plug was observed and hence, the fertilization is assumed to have occurred was set as embryonic day 0.5 (E 0.5). The experiments were executed in embryos at E13.5 and E15.5 in the dorsolateral and medial neocortex, respectively.

The experimental procedures performed in this study were conducted and designed in agreement with the German Animal Welfare Legislation.

9.6 In utero electroporation

The in utero electroporation (IUE) was performed like previously described (Paridaen et al., 2013). Before surgery, pregnant wildtype mice were anesthetized using isoflurane. The pain perception was additionally reduced through subcutaneous injection of the analgesic drug metamizole.

In the following, the peritoneal cavity of the mother surgically was opened and the uterus containing the embryos exposed. The embryos were then intraventricularly injected with 0.5 µg / µl pCAGGS – *RFP* and either 1 µg / µl pCAGGS – empty or 1 µg / µl of pCAGGS – *FAM72A* or FAM72D in 1x PBS containing 0.1% fast green (Sigma, to visualize the intraventricular injection). The electroporations were conducted with 2 x 4 50 ms pulses of 30 V at 1 s intervals. Finally, embryos were harvested 24 h, 48 h (IUE at E13.5, dorsolateral neocortex) or 72 h (IUE at E15.5, medial neocortex) post-electroporation. The post-surgery analgesic treatment was maintained through metamizole containing drinking water for two days (1.33 mg/ml drinking water).

9.7   Fixation and cryosections

After dissection, the embryonic brains were fixated in 4% PFA (paraformaldehyde) overnight at 4 °C mainly to stabilize the tissue architecture, inactivate proteolytic enzymes, increase the resistance of the sample to withstand further processing and staining and to protect the tissue against microbial contamination. Subsequently, the brains were washed in 1x PBS and kept in 30% sucrose solution for 1 more night at 4 °C to achieve a cryoprotection before it was embedded in O.C.T. Finally, cryosections of 14-20 µm thickness were produced using a cryostat with object and knife temperatures from -16 °C to – 22 °C.

9.8   Immunofluorescence and antibodies

The electroporated cryosections were further processed for immunofluorescence stainings in a two days lasting protocol. Firstly, the slides were washed for 10 min in 1x PBS. Afterward, an antigen retrieval was performed using 0.01 M citrate buffer, where the slides were kept in for 1 h at 70 °C and cooled down after that for about 15 min at room temperature (RT). Then, the slides were treated with TritonX100 0.3% to permeabilize the cell membrane for 20 min, before incubation of 25 min in 0.1 M glycine in 1x PBS at RT. Lastly, the slides were washed at least twice with TX buffer (0.2% gelatin, 300 mM NaCl and TritonX100 0,3%). for 20 min in total at RT, before the primary antibody incubation was started at 4 °C overnight. The antibodies were previously diluted in TX buffer. At day 2, slides were washed twice in 1x PBS, and subsequently, TX buffer (30 min, RT) and the secondary antibodies incubated for 1.5 h at RT. Finally, sections were mounted in Mowiol 4-88 and protected using a 170 µm thick cover glass.

The primary and secondary antibodies used for the experiments presented in this book are listed in Tables 19 and 20, respectively.

Table 19: Primary antibodies

1st Antibody	Species	Source	Dilution
phospho Histone 3 (PH3)	rat	abcam (ab10543)	1:300
Ki67	rabbit	abcam (ab15580)	1:300
Sox2	goat	Santa Cruz (sc-17320)	1:300
Tbr2	rabbit	abcam (ab23345)	1:300
Tbr2	mouse	MPI-CBG Antibody Facility	1:400
Olig2	mouse	Millipore (MABN50)	1:300
RFP	rat	ChromoTek (5F8)	1:1000
RFP	rabbit	Rockland antibodies (600-401-379)	1:2000
GFP	rabbit	abcam (ab290)	1:500
GFP	chicken	Aves labs (GFP – 1020)	1:500
FAM72A/B/C/D – sc -240433	goat	Santa Cruz	1:25; 1:40; 1:50
FAM72A – abx145553	rabbit	abbexa	1:40; 1:50

Table 20: Secondary antibodies

2nd Antibody	Source	Dilution
goat anti-rat, Alexa 647	Thermo Fisher (Invitrogen)	1:550
goat anti-rat, Alexa 555	Thermo Fisher (Invitrogen)	1:550
goat anti-rat, Alexa 488	Thermo Fisher (Invitrogen)	1:550
donkey anti - rat, Cy5	Jackson ImmunoResearch	1:550
donkey anti - rabbit, Alexa 647	Thermo Fisher (Invitrogen)	1:550
donkey anti-rabbit Alexa 555	Thermo Fisher (Invitrogen)	1:550
donkey anti - rabbit, Alexa 488	Thermo Fisher (Invitrogen)	1:550
donkey anti-goat, Alexa 647	Thermo Fisher (Invitrogen)	1:550
donkey anti - goat, Cy5	Jackson ImmunoResearch	1:500
donkey anti - mouse, Alexa 555	Thermo Fisher (Invitrogen)	1:550
donkey anti - mouse, Alexa 488	Thermo Fisher (Invitrogen)	1:550
donkey anti - chicken, Alexa 488	Thermo Fisher (Invitrogen)	1:550
donkey anti - chicken, Cy2	Jackson ImmunoResearch	1:550

9.9   5-Ethynyl-2´-deoxyuridine (EdU) detection

For the experiments conducted to analyze the cell-cycle reentry of the FAM72A/D electroporated cells, the intraabdominally injected EdU (see: Results II) had to be detected in addition to the Ki67 immunofluorescence staining. The protocol I used for the first part was identical to the procedure described for the immunofluorescence staining, albeit a few additional steps were performed after the secondary antibody incubation. Firstly, the secondary antibodies were washed with TX buffer. Secondly, the sections were fixated in 4 % PFA at RT for 20 min. Thirdly, the PFA was washed away using 1x PBS and the sections incubated in 0.1 M glycine for 15 min and then, 3 x 5 min in 3 % BSA (bovine serum albumin) in 1x PBS. Finally, the EdU detection solution (387 µl H_2O, 43 µl 10x reaction buffer + 50 µl 10x reaction additive + 20 µl $CuSO_4$ + 1.2 µl Alexa fluor azide) provided by Click-IT EdU Alexa fluor 647 imaging kit (Thermo Fisher, Invitrogen) was added and incubated for 30 min followed by another washing step in 3% BSA and 1x PBS and the subsequent mounting in Mowiol and protection with a 170 µm (thickness) cover glass.

9.10   Image acquisition

The confocal images were obtained using an LSM 510 (Zeiss) confocal microscope. Stacks of typically 1024 x 1024 pixels x 7-20 1 µm optical sections were acquired.

Afterward, the images were analyzed and further processed using IMAGEJ: https://imagej.nih.gov/ij/.

9.11   Statistics

The determined cell counts were analyzed and visualized using Excel (Microsoft) and Prism (GraphPad Software). If only two parameters such as control and FAM72D were compared a Mann-Whitney test with two-tailed p-values was used (two groups of observations that do not follow a normal distribution). In the case of experiments consisting of control, FAM72A and FAM72D datapoints a Kruskal-Wallis test was applied.

9.12   Microdissection and single cell suspension

The electroporated dorsolateral developing mouse neocortex was microdissected under an epifluorescence stereomicroscope to increase the density of electroporated GFP positive cells in the cell suspension made out of the microdissected tissue.

The cortices from the same condition (pCAGGS-empty, pCAGGS-FAM72A or pCAGGS-FAM72D electroporated embryos of the same mother) were pooled together in a 2 ml microcentrifuge tube. Hence, three conditions were processed for each of the four independent experiments. The single cell suspension was produced using the MACS Miltenyi Biotec Neural Tissue Dissociation Kit with an optimized protocol (Florio et al., 2015): 975 µl of solution 2 and 25 µl of the enzyme P (papain) were mixed, added to the brains (max. 4 per sample) of each condition and kept revolving at 37 °C for 15 min. Then, 5 µl per sample of enzyme A were added as well as 10 µl per sample of buffer Y (MACS Miltenyi Biotec Neural Tissue Dissociation Kit). Afterward, the tissue was mechanically demolished using a pipette with filtered tip.

9.13 FACS

To isolate the GFP positive (electroporated) cells from the single cell suspension fluorescence-activated cell sorting (FACS) was used. Therefore, we used a 5-laser – FACSAria Fusion (BD Bioscience) and the FACS Diva software (BD Bioscience v.8.0.1) for the analysis. The cell suspension was filtered through a filter with a 20 µm pore size (Cup filcons, BD Bioscience) to reduce the amounts of cell clumps already in advance and kept on ice until the FACS started. The FACS was performed at 4 °C. The P1 gate was set on the SSC/FSC dot-plot to distinguish living from apoptotic cells based on size and shape which was further restricted by a P2 gate set on the FSC-H/FSC-W to exclude duplets and cell clumps which were not filtered out before. Out of the remaining population, a P3 gate was set using the FITC-A laser to isolate the GFP positive cells.

5000 GFP+ cells per condition were sorted into 300 µl RLT beta-mercaptoethanol buffer (10 µl beta-mercaptoethanol per 1 ml RLT buffer (Qiagen) and shortly vortexed, directly put on dry ice and stored at -80 °C. The quality of the FACS was assessed reanalyzing already sorted GFP positive cells revealing a purity of 90-93%.

9.14 RNA sequencing

The total RNA was extracted using the RNAeasy Mini Kit (Qiagen) according to the protocol provided by the manufacturer and eluted in 20 µl RNAse free water and stored at -80 °C, until it was further processed at the Deep Sequencing Facility at the BIOTEC Dresden under direction from Andreas Dahl. The cDNA was synthesized from 5 µl of the isolated mRNA with the SmartScribe reverse transcriptase (Clontech) using template switching oligos and a universally tailed poly-dT primer. The following amplification of the purified cDNA was performed in 12 PCR cycles using the Advantage 2 DNA polymerase and cleaned up with XP Beads (Beckmann). Afterward, the amplified cDNA was ultrasonic sheared (Covaris LE220) and the samples subjected to standard Illumina fragment library preparation using New England Biolabs Next chemistries.

In a nutshell: fragments were end repaired, A-tailed and ligated to universal Illumina adapters. NGS libraries were indexed and finalized by PCR amplification with indexed PCR primers (15 cycles). The libraries were purified using XP beads (Beckman Coulter) and quantified by qPCR (KAPA Biosystems) and finally subjected to Illumina 75-bp single-end sequencing on the Illumina HiSeq 2500 platform providing an average of 75 Mio reads per sample.

9.15 Transcriptome analysis

The reads were checked for their overall quality using FastQC (v0.11.2). Alignments of reads against the mouse genome reference assembly GRCm38 and quantification of genes of the Ensembl release v81 were done using STAR (v2.5.2b). Differential gene expression analysis on raw counts was performed with DESeq2 (v1.16.1) using a cutoff of $p < 0.01$. The resulting sets differentially expressed genes were tested for enrichment in GO terms and pathways (KEGG, Reactome) on a 1% q-value level. The analysis of the raw data was performed by Holger Brandl (MPI CBG/CSBD) and Marta Florio (MPI CBG).

10 References

Aaku-Saraste E, Hellwig A, Huttner, WB. 1996. Loss of occludin and functional tight junctions, but not ZO-1, during neural tube closure - remodeling of the neuroepithelium prior to neurogenesis. Dev Biol, 180, 664-679.

Abarca-Heidemann K, Friederichs S, Klamp T, Boehm U, Guethlein LA, Ortmann B. 2002. Regulation of the expression of mouse TAP-associated glycoprotein (tapasin) by cytokines. Immunol Lett, 83(3), 197-207.

Aiello, L. C. 1997. Brains and guts in human evolution: The Expensive Tissue Hypothesis. Braz. J. Genet., 20(1), 141-148. doi:10.1590/S0100-84551997000100023.

Amado-Azevedo J, Reinhard NR, van Bezu J, de Menezes RX, van Beusechem VW, van Nieuw Amerongen GP, van Hinsbergh VWM, Hordijk PL. 2017. A CDC42-centered signaling unit is a dominant positive regulator of endothelial integrity. *Sci Rep, 7*(1), 10132. doi:10.1038/s41598-017-10392-0.

Ang YS, Tsai SY, Lee DF, Monk J, Su J, Ratnakumar K, Ding J, Ge Y, Darr H, Chang B, Wang J, Rendl M, Bernstein E, Schaniel C, Lemischka IR. 2011. Wdr5 mediates self-renewal and reprogramming via the embryonic stem cell core transcriptional network. Cell, 145(2), 183-197. doi:10.1016/j.cell.2011.03.003.

Arai Y, Pulvers JN, Haffner C, Schilling B, Nüsslein I, Calegari F, Huttner WB. 2011. Neural stem and progenitor cells shorten S-phase on commitment to neuron production. Nat Commun, 2(1), 154. doi:10.1038/ncomms1155.

Aruga J, Mikoshiba K. 2003. Identification and characterization of Slitrk, a novel neuronal transmembrane protein family controlling neurite outgrowth. Mol Cell Neurosci, 24(1), 117-129. doi:10.1016/S1044-7431(03)00129-5.

Aruga J, Yokota N, Mikoshiba K. 2003. Human SLITRK family genes: genomic organization and expression profiling in normal brain and brain tumor tissue. Gene, 315, 87-94. doi:10.1016/S0378-1119(03)00715-7.

Asenjo AB, Rim J, Oprian DD. 1994. Molecular determinants of human red/green color discrimination. Neuron, 12(5), 1131-1138.

Aston C, Jiang L, Sokolov, BP. 2005. Transcriptional profiling reveals evidence for signaling and oligodendroglial abnormalities in the temporal cortex from patients with major depressive disorder. Mol Psychiatry, 10(3), 309-322. doi:10.1038/sj.mp.4001565.

Azarashvili T, Krestinina O, Galvita A, Grachev D, Baburina Y, Stricker R, Evtodienko Y, Reiser G. 2009. Ca^{2+}- dependent permeability transition regulation in rat brain mitochondria by 2',3'-cyclic nucleotides and 2',3'-cyclic nucleotide 3'-phosphodiesterase. Am J Physiol Cell Physiol, 296(6), C1428-C1439. doi:10.1152/ajpcell.00006.2009.

Azevedo FA, Carvalho LR, Grinberg LT, Farfel JM, Ferretti RE, Leite RE, Jacob Filho W, Lent R, Herculano-Houzel S. 2009. Equal numbers of neuronal and nonneuronal cells make the human brain an isometrically scaled-up primate brain. J Comp Neurol, 513(5), 532-541. doi:10.1002/cne.21974.

Azevedo FAC, Carvalho LRB, Grinberg LT, Farfel JM, Ferretti REL, Leite REP, Jacob Filho W, Lent R, Herculano-Houzel S. 2009. Equal numbers of neuronal and nonneuronal cells make the human brain an isometrically scaled-up primate brain. J Comp Neurol, 513(5), 532-541. doi:10.1002/cne.21974.

Bahram S, Bresnahan M, Geraghty DE, Spies T. 1994. A second lineage of mammalian major histocompatibility complex class I genes. Proc Natl Acad Sci U S A, 91(14), 6259-6263. doi:10.1073/pnas.1714341115.

Bailey JA, Gu Z, Clark RA, Reinert K, Samonte RV, Schwartz S, Adams MD, Myers EW, Li PW, Eichler EE. 2002. Recent segmental duplications in the human genome. Science, 297(5583), 1003-1007. doi:10.1126/science.1072047.

Bailey JA, Liu G, Eichler EE. 2003. An Alu transposition model for the origin and expansion of human segmental duplications. Am J Hum Genet, 73(4), 823-834. doi:10.1086/378594.

Bailey JA, Yavor AM, Massa HF, Trask BJ, Eichler EE. 2001. Segmental duplications: organization and impact within the current human genome project assembly. Genome Res, 11(5), 653-656. doi:10.1101/gr.188901.

Barnabe-Heider F, Miller FD. 2003. Endogenously produced neurotrophins regulate survival and differentiation of cortical progenitors via distinct signaling pathways. J Neurosci, 23(12), 5149-5160.

Barrett L, Henzi P, Rendall D. 2007. Social brains, simple minds: does social complexity really require cognitive complexity? Philos Trans R Soc Lond B Biol Sci, 362(1480), 561-575. doi:10.1098/rstb.2006.1995.

Barski A, Cuddapah S, Cui K, Roh TY, Schones DE, Wang Z, Wei G, Chepelev I, Zhao K. 2007. High-resolution profiling of histone methylations in the human genome. Cell, 129(4), 823-837. doi:10.1016/j.cell.2007.05.009.

Bartkowska K, Paquin A, Gauthier AS, Kaplan DR, Miller, FD. 2007. Trk signaling regulates neural precursor cell proliferation and differentiation during cortical development. Development, 134(24), 4369-4380. doi:10.1242/dev.008227.

Bartosz G. 2009. Reactive oxygen species: destroyers or messengers? Biochem Pharmacol, 77(8), 1303-1315. doi:10.1016/j.bcp.2008.11.009.

Beaubien F, Cloutier JF. 2009. Differential expression of slitrk family members in the mouse nervous system. Dev Dyn, 238(12), 3285-3296. doi:10.1002/dvdy.22160.

Behrend L, Henderson G, Zwacka RM. 2003. Reactive oxygen species in oncogenic transformation. Biochem Soc Trans, 31(Pt 6), 1441-1444. doi:10.1042/.

Benavides-Piccione R, Ballesteros-Yanez I, DeFelipe J, Yuste R. 2002. Cortical area and species differences in dendritic spine morphology. J Neurocytol, 31(3-5), 337-346.

Benayoun BA, Pollina EA, Ucar D, Mahmoudi S, Karra K, Wong ED, Devarajan K, Daugherty AC, Kundaje AB, Mancini E, Hitz BC, Gupta R, Rando TA, Baker JC, Snyder MP, Cherry JM, Brunet A. 2014. H3K4me3 breadth is linked to cell identity and transcriptional consistency. Cell, 158(3), 673-688. doi:10.1016/j.cell.2014.06.027.

Betizeau M, Cortay V, Patti D, Pfister S, Gautier E, Bellemin-Ménard A, Afanassieff M, Huissoud C, Douglas R J, Kennedy H, Dehay C. 2013. Precursor diversity and complexity of lineage relationships in the outer subventricular zone of the primate. Neuron, 80(2), 442-457. doi:10.1016/j.neuron.2013.09.032.

Bifulco M, Laezza C, Stingo S, Wolff J. 2002. 2′,3′-Cyclic nucleotide 3′-phosphodiesterase: A membrane-bound, microtubule-associated protein and membrane anchor for tubulin. Proc Natl Acad Sci U S A, 99(4), 1807-1812. doi:10.1073/pnas.042678799.

Blaschke AJ, Staley K, Chun J. 1996. Widespread programmed cell death in proliferative and postmitotic regions of the fetal cerebral cortex. Development, 122(4), 1165-1174.

Blees A, Januliene D, Hofmann T, Koller N, Schmidt C, Trowitzsch S, Moeller A, Tampé R. 2017. Structure of the human MHC-I peptide-loading complex. Nature, 551(7681), 525. doi:10.1038/nature24627.

Borrell V, Reillo I. 2012. Emerging roles of neural stem cells in cerebral cortex development and evolution. Dev Neurobiol, 72(7), 955-971. doi:10.1002/dneu.22013.

Boudko SP, Ishikawa Y, Nix J, Chapman MS, Bächinger HP. 2014. Structure of human peptidyl-prolyl cis-trans isomerase FKBP22 containing two EF-hand motifs. Protein Sci, 23(1), 67-75. doi:10.1002/pro.2391.

Bourdon JC, Renzing J, Robertson PL, Fernandes KN, Lane DP. 2002. Scotin, a novel p53-inducible proapoptotic protein located in the ER and the nuclear membrane. J Cell Biol, 158(2), 235-246. doi:10.1083/jcb.200203006.

Braun PE, Sandillon F, Edwards A, Matthieu JM, Privat A. 1988. Immunocytochemical localization by electron microscopy of 2'3'-cyclic nucleotide 3'-phosphodiesterase in developing oligodendrocytes of normal and mutant brain. J Neurosci, 8(8), 3057-3066.

Briscoe SD, Ragsdale CW. 2018. Homology, neocortex, and the evolution of developmental mechanisms. Science, 362, 190-193.

Buckner RL, Krienen FM. 2013. The evolution of distributed association networks in the human brain. Trends Cogn Sci, 17(12), 648-665. doi:10.1016/j.tics.2013.09.017.

Buckner RL, Krienen FM. 2017. Human Association Cortex: Expanded, Untethered, Neotenous, and Plastic. In J. H. Kaas (Ed.), Evolution of Nervous System (2nd ed., Vol. 4, pp. 169-182). Elsevier, Amsterdam.

Calegari F, Haubensak W, Haffner C, Huttner WB. 2005. Selective lengthening of the cell cycle in the neurogenic subpopulation of neural progenitor cells during mouse brain development. J Neurosci, 25(28), 6533-6538.

Calegari F, Huttner WB. 2003. An inhibition of cyclin-dependent kinases that lengthens, but does not arrest, neuroepithelial cell cycle induces premature neurogenesis. J Cell Sci, 116(Pt 24), 4947-4955.

Calkins MJ, Manczak M, Mao P, Shirendeb U, Reddy PH. 2011. Impaired mitochondrial biogenesis, defective axonal transport of mitochondria, abnormal mitochondrial dynamics and synaptic degeneration in a mouse model of Alzheimer's disease. Hum Mol Genet, 20(23), 4515-4529. doi:10.1093/hmg/ddr381.

Campbell K, Götz M. (2002). Radial glia: multi-purpose cells for vertebrate brain development. Trends Neurosci, 25(5), 235-238.

Caviness VS Jr, Takahashi T, Nowakowski RS. 1995. Numbers, time and neocortical neuronogenesis: a general developmental and evolutionary model. Trends Neurosci, 18(9), 379-383. doi:10.1016/0166-2236(95)93933-O.

Caviness VS Jr, Takahashi T, Nowakowski RS. 1995. Numbers, time and neocortical neuronogenesis: a general developmental and evolutionary model. Trends Neurosci, 18(9), 379-383.

Charrier C, Joshi K, Coutinho-Budd J, Kim JE, Lambert N, de Marchena J, Jin WL, Vanderhaeghen P, Ghosh A, Sassa T, Polleux F. (2012). Inhibition of SRGAP2 Function by Its Human-Specific Paralogs Induces Neoteny during Spine Maturation. Cell, 149(4), 923-935. doi:10.1016/j.cell.2012.03.034.

Chen FC, Vallender EJ, Wang H, Tzeng CS, Li WH. 2001. Genomic divergence between human and chimpanzee estimated from large-scale alignments of genomic sequences. J Hered, 92(6), 481-489.

Chen JM, Chuzhanova N, Stenson PD, Férec C, Cooper DN. (2005). Meta-Analysis of gross insertions causing human genetic disease: Novel mutational mechanisms and the role of replication slippage. Hum Mutat, 25(2), 207-221. doi:10.1002/humu.20133.

Chen L, DeVries AL, Cheng CH. 1997. Evolution of antifreeze glycoprotein gene from a trypsinogen gene in Antarctic notothenioid fish. Proc Natl Acad Sci U S A, 94(8), 3811-3816.

Chenn A, Zhang YA, Chang BT, McConnell SK. 1998. Intrinsic polarity of mammalian neuroepithelial cells. Mol Cell Neurosci, 11(4), 183-193.

Cho IT, Lim Y, Golden JA, Cho G. 2017. Aristaless Related Homeobox (ARX) Interacts with β-Catenin, BCL9, and P300 to Regulate Canonical Wnt Signaling. PLoS One, 12(1). doi:10.1371/journal.pone.0170282.

Ciciotte SL, Gwynn B, Moriyama K, Huizing M, Gahl WA, Bonifacino JS, Peters LL. 2003. Cappuccino, a mouse model of Hermansky-Pudlak syndrome, encodes a novel protein that is part of the pallidin-muted complex (BLOC-1). Blood, 101(11), 4402-4407. doi:10.1182/blood-2003-01-0020.

Coman D, Fullston T, Shoubridge C, Leventer R, Wong F, Nazaretian S, Simpson I, Gecz J, McGillivray G. 2017. X-Linked Lissencephaly With Absent Corpus Callosum and Abnormal Genitalia: An Evolving Multisystem Syndrome With Severe Congenital Intestinal Diarrhea Disease. Child Neurol Open, 4(5). doi:10.1177/2329048X17738625.

Conant GC, Wolfe KH. 2008. Turning a hobby into a job: how duplicated genes find new functions. Nat Rev Genet, 9(12), 938-950. doi:10.1038/nrg2482.

Cornélio AM, de Bittencourt-Navarrete RE, de Bittencourt Brum R, Queiroz CM, Costa MR. 2016. Human Brain Expansion during Evolution Is Independent of Fire Control and Cooking. Front Neurosci, 10, 199. doi:10.3389/fnins.2016.00167.

Court H, Ahearn IM, Amoyel M, Bach EA, Philips MR. 2017. Regulation of NOTCH signaling by RAB7 and RAB8 requires carboxyl methylation by ICMT. J Cell Biol, 216(12), 4165-4182. doi:10.1083/jcb.201701053.

Cuylen S, Blaukopf C, Politi AZ, Müller-Reichert T, Neumann B, Poser I, Ellenberg J, Hyman AA, Gerlich DW. 2016. Ki-67 acts as a biological surfactant to disperse mitotic chromosomes. Nature, 535(7611), 308-312. doi:10.1038/nature18610.

Darwin, C. (1871). The descent of man and selection in relation to sex.1. Murray, London.

Darwin C. (1959). On the origin of species by means of natural selection. Penguin Classics, London.

de Monasterio-Schrader P, Jahn O, Tenzer S, Wichert SP, Patzig J, Werner HB. 2012. Systematic approaches to central nervous system myelin. Cell Mol Life Sci, 69(17), 2879-2894. doi:10.1007/s00018-012-0958-9.

Dean EJ, Davis JC, Davis RW, Petrov DA. 2008. Pervasive and persistent redundancy among duplicated genes in yeast. PLoS genetics, 4(7). doi:10.1371/journal.pgen.1000113.

Deaner RO, Isler K, Burkart J, van Schaik C. 2007. Overall Brain Size, and Not Encephalization Quotient, Best Predicts Cognitive Ability across Non-Human Primates. Brain Behav Evol, 70(2), 115-124. doi:10.1159/000102973.

Dehay C, Kennedy H, Kosik KS. 2015. The outer subventricular zone and primate-specific cortical complexification. Neuron, 85(4), 683-694. doi:10.1016/j.neuron.2014.12.060.

Dennis MY, Eichler EE. 2016. Human adaptation and evolution by segmental duplication. Curr Opin Genet Dev, 41, 44-52. doi:10.1016/j.gde.2016.08.001.

Dennis MY, Nuttle X, Sudmant PH, Antonacci F, Graves TA, Nefedov M, Rosenfeld JA, Sajjadian S, Malig M, Kotkiewicz H, Curry CJ, Shafer S, Shaffer LG, de Jong PJ, Wilson RK, Eichler EE. 2012. Evolution of human-specific neural SRGAP2 genes by incomplete segmental duplication. Cell, 149(4), 912-922. doi:10.1016/j.cell.2012.03.033.

Detmer SA, Chan DC. 2007. Functions and dysfunctions of mitochondrial dynamics. Nat Rev Mol Cell Biol, 8(11), 870-879. doi:10.1038/nrm2275.

Di Lullo E, Kriegstein AR. 2017. The use of brain organoids to investigate neural development and disease. Nat Rev Neurosci, 18(10), 573-584. doi:10.1038/nrn.2017.107.

Diamond IT, Hall WC. 1969. Evolution of neocortex. Science, 164(3877), 251-262.

Dissemond J, Kothen T, Mörs J, Weimann TK, Lindeke A, Goos M, Wagner SN. 2003. Downregulation of tapasin expression in progressive human malignant melanoma. Arch Dermatol Res, 295(2), 43-49. doi:10.1007/s00403-003-0393-8.

Douglas AJ, Fox MF, Abbott CM, Hinks LJ, Sharpe G, Povey S, Thompson RJ. 1992. Structure and chromosomal localization of the human 2',3'-cyclic nucleotide 3'-phosphodiesterase gene. Ann Hum Genet, 243-254.

Draeby I, Woods YL, la Cour JM, Mollerup J, Bourdon JC, Berchtold MW. 2007. The calcium binding protein ALG-2 binds and stabilizes Scotin, a p53-inducible gene product localized at the endoplasmic reticulum membrane. Arch Biochem Biophys, 467(1), 87-94. doi:10.1016/j.abb.2007.07.028.

Drummond DA, Raval A, Wilke CO. 2006. A Single Determinant Dominates the Rate of Yeast Protein Evolution. Mol Biol Evol, 23(2), 327-337. doi:10.1093/molbev/msj038.

Duff BJ, Macritchie KAN, Moorhead TWJ, Lawrie SM, Blackwood DHR. 2013. Human brain imaging studies of DISC1 in schizophrenia, bipolar disorder and depression: a systematic review. Schizophr Res, 147(1), 1-13. doi:10.1016/j.schres.2013.03.015.

Dunbar RI. 2009. The social brain hypothesis and its implications for social evolution. Ann Hum Biol, 36(5), 562-572. doi:10.1080/03014460902960289.

Dunbar RIM, Shultz S. 2007. Understanding primate brain evolution. Philos Trans R Soc Lond B Biol Sci, 362(1480), 649-658. doi:10.1098/rstb.2006.2001.

Dunbar RIM, Shultz S. 2017. Why are there so many explanations for primate brain evolution? Philos Trans R Soc Lond B Biol Sci, 372(1727), 20160244. doi:10.1098/rstb.2016.0244.

Englund C, Fink A, Lau C, Pham D, Daza RA, Bulfone A, Kowalczyk T, Hevner R. F. 2005. Pax6, Tbr2, and Tbr1 are expressed sequentially by radial glia, intermediate progenitor cells, and postmitotic neurons in developing neocortex. J Neurosci, 25(1), 247-251.

Englund C, Fink A, Lau C, Pham D, Daza RAM, Bulfone A, Kowalczyk T, Hevner RF. 2005. Pax6, Tbr2, and Tbr1 Are Expressed Sequentially by Radial Glia, Intermediate Progenitor Cells, and Postmitotic Neurons in Developing Neocortex. J Neurosci, 25(1), 247-251. doi:10.1523/JNEUROSCI.2899-04.2005.

Estivill-Torrus G, Pearson H, van Heyningen V, Price DJ, Rashbass P. 2002. Pax6 is required to regulate the cell cycle and the rate of progression from symmetrical to asymmetrical division in mammalian cortical progenitors. Development, 129(2), 455-466.

Famulski JK, Trivedi N, Howell D, Yang Y, Tong Y, Gilbertson R, Solecki DJ. 2010. Siah regulation of Pard3A controls neuronal cell adhesion during germinal zone exit. Science, 330(6012), 1834-1838. doi:10.1126/science.1198480.

Fietz SA, Kelava I, Vogt J, Wilsch-Bräuninger M, Stenzel D, Fish JL, Corbeil D, Riehn A, Distler W, Nitsch R, Huttner WB. 2010. OSVZ progenitors of human and ferret neocortex are epithelial-like and expand by integrin signaling. Nat Neurosci, 13(6), 690-699. doi:10.1038/nn.2553.

Fietz SA, Lachmann R, Brandl H, Kircher M, Samusik N, Schröder R, Lakshmanaperumal N, Henry I, Vogt J, Riehn A, Distler W, Nitsch R, Enard W, Pääbo S, Huttner WB. 2012. Transcriptomes of germinal zones of human and mouse fetal neocortex suggest a role of extracellular matrix in progenitor self-renewal. Proc Natl Acad Sci U S A, 109(29), 11836-11841. doi:10.1073/pnas.1209647109.

Florio M. 2015. Transcriptional Signatures of Neocortical Expansion - Evolutionary Role of the Human-Specific Gene ARHGAP11B. Technische Universität Dresden, Ph.D.

Florio M, Albert M, Taverna E, Namba T, Brandl H, Lewitus E, Haffner C, Sykes A, Wong FK, Peters J, Guhr E, Klemroth S, Prüfer K, Kelso J, Naumann R, Nüsslein I, Dahl A, Lachmann R, Pääbo S, Huttner WB. 2015. Human-specific gene ARHGAP11B promotes basal progenitor amplification and neocortex expansion. Science, 347(6229), 1465-1470. doi:10.1126/science.aaa1975.

Florio M, Borrell V, Huttner WB. 2017. Human-specific genomic signatures of neocortical expansion. Curr Opin Neurobiol, 42, 33-44. doi:10.1016/j.conb.2016.11.004.

Florio M, Heide M, Pinson A, Brandl H, Albert M, Winkler S, Wimberger P, Huttner WB, Hiller M. 2018. Evolution and cell-type specificity of human-specific genes preferentially expressed in progenitors of fetal neocortex. eLife, 7. doi:10.7554/eLife.32332.

Florio M, Huttner WB. 2014. Neural progenitors, neurogenesis and the evolution of the neocortex. Development, 141(11), 2182-2194. doi:10.1242/dev.090571.

Florio M, Namba T, Pääbo S, Hiller M, Huttner WB. 2016. A single splice site mutation in human-specific ARHGAP11B causes basal progenitor amplification. Sci Adv, 2(12). doi:10.1126/sciadv.1601941.

Fonseca-Azevedo K, Herculano-Houzel S. 2012. Metabolic constraint imposes tradeoff between body size and number of brain neurons in human evolution. Proc Natl Acad Sci U S A, 109(45), 18571-18576. doi:10.1073/pnas.1206390109.

Fontdevila A. 2011. The Dynamic Genome: A Darwinian Approach. University Press, Oxford.

Force A, Lynch M, Pickett FB, Amores A, Yan YL, Postlethwait J. 1999. Preservation of duplicate genes by complementary, degenerative mutations. Genetics, 151(4), 1531-1545.

Fortna A, Kim Y, MacLaren E, Marshall K, Hahn G, Meltesen L, Brenton M, Hink R, Burgers S, Hernandez-Boussard T, Karimpour-Fard A, Glueck D, McGavran L, Berry R, Pollack J, Sikela JM. 2004. Lineage-specific gene duplication and loss in human and great ape evolution. PLoS Biol, 2(7), E207. doi:10.1371/journal.pbio.0020207.

Fossati M, Pizzarelli R, Schmidt ER, Kupferman JV, Stroebel D, Polleux F, Charrier C. 2016. SRGAP2 and Its Human-Specific Paralog Co-Regulate the Development of Excitatory and Inhibitory Synapses. Neuron, 91(2), 356-369. doi:10.1016/j.neuron.2016.06.013.

Friedman R, Hughes AL. 2003. The temporal distribution of gene duplication events in a set of highly conserved human gene families. Mol Biol Evol, 20(1), 154-161.

Fuchs JL, Schwark HD. 2004. Neuronal primary cilia: a review. Cell Biol Int, 28(2), 111-118. doi:10.1016/j.cellbi.2003.11.008.

Gal JS, Morozov YM, Ayoub AE, Chatterjee M, Rakic P, Haydar TF. 2006. Molecular and morphological heterogeneity of neural precursors in the mouse neocortical proliferative zones. J Neurosci, 26(3), 1045-1056.

Garlanda C, Bottazzi B, Salvatori G, De Santis R, Cotena A, Deban L, Maina V, Moalli F, Doni A, Veliz-Rodriguez T, Mantovani A. 2006. Pentraxins in innate immunity and inflammation. Novartis Found Symp, 279, 80-86- discussion 86-91- 216-219.

Gaur RK. 2014. Amino acid frequency distribution among eukaryotic proteins. IIOAB J, 5(2), 6-11.

Gertz CC, Lui JH, Lamonica BE, Wang X, Kriegstein AR. 2014. Diverse behaviors of outer radial glia in developing ferret and human cortex. J Neurosci, 34(7), 2559-2570. doi:10.1523/JNEUROSCI.2645-13.2014.

Gilad Y, Man O, Pääbo S, Lancet D. 2003. Human specific loss of olfactory receptor genes. Proc Natl Acad Sci U S A, 100(6), 3324-3327. doi:10.1073/pnas.0535697100.

Glazko GV, Nei M. 2003. Estimation of divergence times for major lineages of primate species. Mol Biol Evol, 20(3), 424-434. doi:10.1093/molbev/msg050.

Götz M, Huttner WB. 2005. The cell biology of neurogenesis. Nat Rev Mol Cell Biol, 6(10), 777-788. doi:10.1038/nrm1739.

Götz M, Stoykova A, Gruss P. 1998. Pax6 controls radial glia differentiation in the cerebral cortex. Neuron, 21(5), 1031-1044.

Goymer P. 2007. Colour vision for mice. Nat Rev Neurosci, 8(5), 328-329. doi:10.1038/nrn2135.

Groh V, Bahram S, Bauer S, Herman A, Beauchamp M, Spies T. 1996. Cell stress-regulated human major histocompatibility complex class I gene expressed in gastrointestinal epithelium. Proc Natl Acad Sci U S A, 93(22), 12445-12450. doi:10.1073/pnas.1714341115.

Gu Z, Nicolae D, Lu HHS, Li WH. 2002. Rapid divergence in expression between duplicate genes inferred from microarray data. Trends Genet, 18(12), 609-613.

Guerrier S, Coutinho-Budd J, Sassa T, Gresset A, Jordan NV, Chen K, Jin WL, Frost A, Polleux F. 2009. The F-BAR domain of srGAP2 induces membrane protrusions required for neuronal migration and morphogenesis. Cell, 138(5), 990-1004. doi:10.1016/j.cell.2009.06.047.

Guo C, Zhang X, Fink SP, Platzer P, Wilson K, Willson JKV, Wang Z, Markowitz SD. 2008. Ugene, a Newly Identified Protein That Is Commonly Overexpressed in Cancer and Binds Uracil DNA Glycosylase. Cancer Res, 68(15), 6118-6126. doi:10.1158/0008-5472.CAN-08-1259.

Guo H, Zhu P, Yan L, Li R, Hu B, Lian Y, Yan J, Ren X, Lin S, Li J, Jin X, Shi X, Liu P, Wang X, Wang W, Wei Y, Li X, Guo F, Wu X, Fan X, Yong J, Wen L, Xie SX, Tang F, Qiao J. 2014. The DNA methylation landscape of human early embryos. Nature, 511, 606. doi:10.1038/nature13544.

Gupta V, You Y, Gupta V, Klistorner A, Graham S. 2013. TrkB Receptor Signalling: Implications in Neurodegenerative, Psychiatric and Proliferative Disorders. Int J Mol Sci, 14(5), 10122-10142. doi:10.3390/ijms140510122.

Hagemeyer N, Goebbels S, Papiol S, Kästner A, Hofer S, Begemann M, Gerwig UC, Boretius S, Wieser GL, Ronnenberg A, Gurvich A, Heckers SH, Frahm J, Nave KA, Ehrenreich H. 2012. A myelin gene causative of a catatonia-depression syndrome upon aging. EMBO Mol Med, 4(6), 528-539. doi:10.1002/emmm.201200230.

Hans F, Dimitrov S. 2001. Histone H3 phosphorylation and cell division. Oncogene, 20(24), 3021-3027. doi:10.1038/sj.onc.1204326.

Hansen DV, Lui JH, Parker PR, Kriegstein AR. 2010. Neurogenic radial glia in the outer subventricular zone of human neocortex. Nature, 464(7288), 554-561. doi:10.1038/nature08845.

Harari YN. 2014. Sapiens: a brief history of humankind. Vintage Books, London.

Hart RP, Goff LA. 2016. Long noncoding RNAs: Central to nervous system development. Int J Dev Neurosci, 55, 109-116. doi:10.1016/j.ijdevneu.2016.06.001.

Hartfuss E, Galli R, Heins N, Götz M. 2001. Characterization of CNS precursor subtypes and radial glia. Dev Biol, 229(1), 15-30.

Harvey PH, Clutton-Brock TH, Mace GM. 1980. Brain size and ecology in small mammals and primates. Proc Natl Acad Sci U S A, 77(7), 4387-4389.

Haubensak W, Attardo A, Denk W, Huttner WB. 2004. Neurons arise in the basal neuroepithelium of the early mammalian telencephalon: A major site of neurogenesis. Proc Natl Acad Sci U S A, 101, 3196-3201.

Havranek T, Lestanova Z, Mravec B, Strbak V, Bakos J, Bacova Z. 2017. Oxytocin Modulates Expression of Neuron and Glial Markers in the Rat Hippocampus. Folia Biol, 63(3), 91-97.

Haydar TF, Kuan CY, Flavell RA, Rakic P. 1999. The Role of Cell Death in Regulating the Size and Shape of the Mammalian Forebrain. Cereb Cortex, 9(6), 621-626. doi:10.1093/cercor/9.6.621.

He X, Zhang J. 2005. Rapid Subfunctionalization Accompanied by Prolonged and Substantial Neofunctionalization in Duplicate Gene Evolution. Genetics, 169(2), 1157-1164. doi:10.1534/genetics.104.037051.

Heese K. 2013. The protein p17 signaling pathways in cancer. Tumour Biology, 34(6), 4081-4087. doi:10.1007/s13277-013-0999-1.

Heide M, Long KR, Huttner WB. 2017. Novel gene function and regulation in neocortex expansion. Curr Opin Cell Biol, 49, 22-30. doi:10.1016/j.ceb.2017.11.008.

Herceg Z, Wang ZQ. 2001. Functions of poly(ADP-ribose) polymerase (PARP) in DNA repair, genomic integrity and cell death. Mutat Res, 477(1-2), 97-110. doi:10.1016/S0027-5107(01)00111-7.

Herculano-Houzel S. (2009). The human brain in numbers: a linearly scaled-up primate brain. Front Hum Neurosci, 3, 31. doi:10.3389/neuro.09.031.2009.

Hinman JD, Chen CD, Oh SY, Hollander W, Abraham CR. 2008. Age-dependent accumulation of ubiquitinated 2',3'-cyclic nucleotide 3'-phosphodiesterase in myelin lipid rafts. Glia, 56(1), 118-133. doi:10.1002/glia.20595.

Ho NTT, Kim PS, Kutzner A, Heese K. 2017. Cognitive Functions: Human vs. Animal – 4:1 Advantage |-FAM72–SRGAP2-|. J Mol Neurosci, 158, 1-4. doi:10.1007/s12031-017-0901-5.

Ho NTT, Kutzner A, Heese K. 2017. Brain plasticity, cognitive functions and neural stem cells: a pivotal role for the brain-specific neural master gene |-SRGAP2-FAM72-|. Biol Chem, 0(0). doi:10.1515/hsz-2017-0190.

Huang ZM, Chinen M, Chang PJ, Xie T, Zhong L, Demetriou S, Patel MP, Scherzer R, Sviderskaya EV, Bennett DC, Millhauser GL, Oh DH, Cleaver JE, Wei ML. 2012. Targeting protein-trafficking pathways alters melanoma treatment sensitivity. Proc Natl Acad Sci U S A, 109(2), 553-558. doi:10.1073/pnas.1118366109.

Hughes AL. 1994. The Evolution of Functionally Novel Proteins after Gene Duplication. Proc R Soc Lond B Biol Sci, 256(1346), 119-124. doi:10.1098/rspb.1994.0058.

Hurles M. 2004. Gene duplication: the genomic trade in spare parts. PLoS Biol, 2(7), E206. doi:10.1371/journal.pbio.0020206.

Huttner WB, Brand M. 1997. Asymmetric division and polarity of neuroepithelial cells. Curr Opin Neurobiol, 7(1), 29-39.

Huttner WB, Kosodo Y. 2005. Symmetric versus asymmetric cell division during neurogenesis in the developing vertebrate central nervous system. Curr Opin Cell Biol, 17(6), 648-657.

Jaendling A, McFarlane RJ. 2010. Biological roles of translin and translin-associated factor-X: RNA metabolism comes to the fore. Biochem J, 429(2), 225-234. doi:10.1042/BJ20100273.

Jerison H. 1985. Animal intelligence as encephalization. Phil Trans R Soc Lond B *308*, 21-35.

Jerison HJ. 1973. Evolution of the brain and intelligence. Academic Press, New York.

Jia CW, Wang L, Lan YL, Song R, Zhou LY, Yu L, Yang Y, Liang Y, Li Y, Ma YM, Wang SY. 2015. Aneuploidy in Early Miscarriage and its Related Factors. Chin Med J, 128(20), 2772-2776. doi:10.4103/0366-6999.167352.

Jobling MA, Hurles ME, Tyler-Smith C. 2004. Human Evolutionary Genetics. Garland Science, New York.

Joseph RM. 2014. Neuronatin gene: Imprinted and misfolded: Studies in Lafora disease, diabetes and cancer may implicate NNAT-aggregates as a common downstream participant in neuronal loss. Genomics, 103(2-3), 183-188. doi:10.1016/j.ygeno.2013.12.001.

Joukov V, Pajusola K, Kaipainen A, Chilov D, Lahtinen I, Kukk E, Saksela O, Kalkkinen N, Alitalo K. 1996. A novel vascular endothelial growth factor, VEGF-C, is a ligand for the Flt4 (VEGFR-3) and KDR (VEGFR-2) receptor tyrosine kinases. EMBO J, 15(2), 290-298. doi:10.1002/j.1460-2075.1996.tb00359.x.

Kaas JH. 2013. The Evolution of Brains from Early Mammals to Humans. Wiley Interdiscip Rev Cogn Sci, 4(1), 33-45. doi:10.1002/wcs.1206.

Kaessmann H. 2010. Origins, evolution, and phenotypic impact of new genes. Genome Res, 20(10), 1313-1326. doi:10.1101/gr.101386.109.

Kalinka AT, Varga KM, Gerrard DT, Preibisch S, Corcoran DL, Jarrells J, Ohler U, Bergman CM,Tomancak P. 2010. Gene expression divergence recapitulates the developmental hourglass model. Nature, 468, 811. doi:10.1038/nature09634.

Kang YS, Seok HJ, Jeong EJ, Kim Y, Yun SJ, Min JK, Kim SJ, Kim JS. 2016. DUSP1 induces paclitaxel resistance through the regulation of p-glycoprotein expression in human ovarian cancer cells. Biochem Biophys Res Commun, 478(1), 403-409. doi:10.1016/j.bbrc.2016.07.035.

Karkkainen MJ, Haiko P, Sainio K, Partanen J, Taipale J, Petrova TV, Jeltsch M, Jackson DG, Talikka M, Rauvala H, Betsholtz C, Alitalo K. 2004. Vascular endothelial growth factor C is required for sprouting of the first lymphatic vessels from embryonic veins. Nat Immunol, 5(1), 74-80. doi:10.1038/ni1013.

Khacho M, Clark A, Svoboda DS, Azzi J, MacLaurin JG, Meghaizel C, Sesaki H, Lagace DC, Germain M, Harper ME, Park DS, Slack RS. 2016. Mitochondrial Dynamics Impacts Stem Cell Identity and Fate Decisions by Regulating a Nuclear Transcriptional Program. Cell Stem Cell, 19(2), 232-247. doi:10.1016/j.stem.2016.04.015.

Khacho M, Clark A, Svoboda DS, MacLaurin JG, Lagace DC, Park DS, Slack RS. 2017. Mitochondrial dysfunction underlies cognitive defects as a result of neural stem cell depletion and impaired neurogenesis. Hum Mol Genet, 26(17), 3327-3341. doi:10.1093/hmg/ddx217.

Khacho M, Slack RS. 2018. Mitochondrial dynamics in the regulation of neurogenesis: From development to the adult brain. Dev Dyn, 247(1), 47-53. doi:10.1002/dvdy.24538.

Kimura M. 1979. The neutral theory of molecular evolution. Sci Am, 241(5), 98-100, 102, 108 passim.

Kimura M, Ohta T. 1974. On some principles governing molecular evolution. Proc Natl Acad Sci U S A, 71(7), 2848-2852.

Kinoshita MN, Reillo I, Artegiani B, Martínez MÁM, Nelson M, Borrell V, Calegari F. 2013. Regulation of cerebral cortex size and folding by expansion of basal progenitors. EMBO J, 32(13), 1817-1828. doi:10.1038/emboj.2013.96.

Knobloch M, Braun SM, Zurkirchen L, von Schoultz C, Zamboni N, Arauzo-Bravo MJ, Kovacs WJ, Karalay O, Suter U, Machado RA, Roccio M, Lutolf MP, Semenkovich CF, Jessberger S. 2013. Metabolic control of adult neural stem cell activity by Fasn-dependent lipogenesis. Nature, 493(7431), 226-230. doi:10.1038/nature11689.

Koszul R, Caburet S, Dujon B, Fischer G. 2004. Eucaryotic genome evolution through the spontaneous duplication of large chromosomal segments. EMBO J, 23(1), 234-243. doi:10.1038/sj.emboj.7600024.

Kowalczyk T, Pontious A, Englund C, Daza RA, Bedogni F, Hodge R, Attardo A, Bell C, Huttner WB, Hevner RF. 2009. Intermediate neuronal progenitors (basal progenitors) produce pyramidal-projection neurons for all layers of cerebral cortex. Cereb Cortex, 19(10), 2439-2450. doi:10.1093/cercor/bhn260.

Kowalczyk T, Pontious A, Englund C, Daza RAM, Bedogni F, Hodge R, Attardo A, Bell C, Huttner WB, Hevner RF. 2009. Intermediate neuronal progenitors (basal progenitors) produce pyramidal-projection neurons for all layers of cerebral cortex. Cereb Cortex, 19(10), 2439-2450. doi:10.1093/cercor/bhn260.

Krestinina O, Azarashvili T, Baburina Y, Galvita A, Grachev D, Stricker R, Reiser G. 2015. In aging, the vulnerability of rat brain mitochondria is enhanced due to reduced level of 2',3'-cyclic nucleotide-3'-phosphodiesterase (CNP) and subsequently increased permeability transition in brain mitochondria in old animals. Neurochem Int, 80, 41-50. doi:10.1016/j.neuint.2014.09.008.

Kriegstein A, Alvarez-Buylla A. 2009. The glial nature of embryonic and adult neural stem cells. Annu Rev Neurosci, 32, 149-184. doi:10.1146/annurev.neuro.051508.135600.

Kriegstein AR, Götz M. 2003. Radial glia diversity: a matter of cell fate. Glia, 43(1), 37-43.

Krupenye C, Kano F, Hirata S, Call J, Tomasello M. 2016. Great apes anticipate that other individuals will act according to false beliefs. Science, 354(6308), 110-114. doi:10.1126/science.aaf8110.

Kuhlwilm M, de Manuel M, Nater A, Greminger MP, Krützen M, Marques-Bonet T. 2016. Evolution and demography of the great apes. Curr Opin Genet Dev, 41, 124-129. doi:10.1016/j.gde.2016.09.005.

Kurihara T, Tsukada Y. 1967. The regional and subcellular distribution of 2',3'-cyclic nucleotide 3'-phosphohydrolase in the central nervous system. J Neurochem, 14(12), 1167-1174. doi:10.1111/j.1471-4159.1967.tb06164.x.

Kursula P. 2006. Structural properties of proteins specific to the myelin sheath. Amino Acids, 34(2), 175-185. doi:10.1007/s00726-006-0479-7.

Kutzner A, Pramanik S, Kim PS, Heese K. 2015. All-or-(N)One – an epistemological characterization of the human tumorigenic neuronal paralogous FAM72 gene loci. Genomics, 106(5), 278-285. doi:10.1016/j.ygeno.2015.07.003.

Lander ES, Linton LM, Birren B, Nusbaum C, Zody MC, Baldwin J, Devon K, Dewar K, Doyle M, FitzHugh W, Funke R, Gage D, Harris K, Heaford A, Howland J, Kann L, Lehoczky J, LeVine R, McEwan P, McKernan K, Meldrim J, Mesirov JP, Miranda C, Morris W, Naylor J, Raymond,

C, Rosetti M, Santos R, Sheridan A, Sougnez C, Stange-Thomann N, Stojanovic N, Subramanian A, Wyman D, Rogers J, Sulston J, Ainscough R, Beck S, Bentley D, Burton J, Clee C, Carter N, Coulson A, Deadman R, Deloukas P, Dunham A, Dunham I, Durbin R, French L, Grafham D, Gregory S, Hubbard T, Humphray S, Hunt A, Jones M, Lloyd C, McMurray A, Matthews L, Mercer S, Milne S, Mullikin JC, Mungall A, Plumb R, Ross M, Shownkeen R, Sims S, Waterston RH, Wilson RK, Hillier LW, McPherson, JD, Marra MA, Mardis ER, Fulton LA, Chinwalla AT, Pepin KH, Gish WR, Chissoe SL, Wendl MC, Delehaunty KD, Miner TL, Delehaunty A, Kramer JB, Cook LL, Fulton RS, Johnson DL, Minx PJ, Clifton SW, Hawkins T, Branscomb E, Predki P, Richardson P, Wenning S, Slezak T, Doggett N, Cheng JF, Olsen A, Lucas S, Elkin C, Uberbacher E, Frazier M, Gibbs RA, Muzny DM, Scherer SE, Bouck JB, Sodergren EJ, Worley KC, AB, Rives, CM, Gorrell JH, Metzker ML, Naylor SL, Kucherlapati RS, Nelson DL, Weinstock GM, Sakaki Y, Fujiyama A, Hattori M, Yada T, Toyoda A, Itoh T, Kawagoe C, Watanabe H, Totoki Y, Taylor T, Weissenbach J, Heilig R, Saurin W, Artiguenave F, Brottier P, Bruls T, Pelletier E, Robert C, Wincker P, Smith DR, Doucette-Stamm L, Rubenfield M, Weinstock K, Lee HM, Dubois J, Rosenthal A, Platzer M, Nyakatura G, Taudien S, Rump A, Yang H, Yu J, Wang J, Huang G, Gu J, Hood L, Rowen L, Madan A, Qin S, Davis RW, Federspiel NA, Abola AP, Proctor MJ, Myers RM, Schmutz J, Dickson M, Grimwood J, Cox DR, Olson MV, Kaul R, Shimizu N, Kawasaki K, Minoshima S, Evans GA, Athanasiou M, Schultz R, Roe BA, Chen F, Pan H, Ramser J, Lehrach H, Reinhardt R, McCombie WR, de la Bastide M, Dedhia N, Blocker H, Hornischer K, Nordsiek G, Agarwala R, Aravind L, Bailey JA, Bateman A, Batzoglou S, Birney E, Bork P, Brown DG, Burge CB, Cerutti L, Chen HC, Church D, Clamp M, Copley RR, Doerks T, Eddy SR, Eichler EE, Furey TS, Galagan J, Gilbert JG, Harmon C, Hayashizaki Y, Haussler D, Hermjakob H, Hokamp K, Jang W, Johnson, LS, Jones TA, Kasif S, Kaspryzk A, Kennedy S, Kent WJ, Kitts P, Koonin EV, Korf I, Kulp D, Lancet D, Lowe TM, McLysaght A, Mikkelsen T, Moran JV, Mulder N, Pollara VJ, Ponting CP, Schuler G, Schultz J, Slater G, Smit AF, Stupka E, Szustakowski J, Thierry-Mieg D, Thierry-Mieg J, Wagner L, Wallis J, Wheeler R, Williams A, Wolf YI, Wolfe KH, Yang SP, Yeh RF, Collins F, Guyer MS, Peterson J, Felsenfeld A, Wetterstrand KA, Patrinos A, Morgan MJ, de Jong P, Catanese JJ, Osoegawa K, Shizuya H, Choi S, Chen YJ. 2001. Initial sequencing and analysis of the human genome. Nature, 409(6822), 860-921. doi:10.1038/35057062.

Lange C, Huttner WB, Calegari F. 2009. Cdk4/cyclinD1 overexpression in neural stem cells shortens G1, delays neurogenesis, and promotes the generation and expansion of basal progenitors. Cell Stem Cell, 5(3), 320-331. doi:10.1016/j.stem.2009.05.026.

Lange C, Turrero García M, Decimo I, Bifari F, Eelen G, Quaegebeur A, Boon R, Zhao H, Boeckx B, Chang J, Wu C, Le Noble F, Lambrechts D, Dewerchin M, Kuo CJ, Huttner WB,

Carmeliet P. 2016. Relief of hypoxia by angiogenesis promotes neural stem cell differentiation by targeting glycolysis. EMBO J, 35(9), 924-941. doi:10.15252/embj.201592372.

Lange S, Perera S, Teh P, Chen J. 2012. Obscurin and KCTD6 regulate cullin-dependent small ankyrin-1 (sAnk1.5) protein turnover. Mol Biol Cell, 23(13), 2490-2504. doi:10.1091/mbc.E12-01-0052.

Lappe-Siefke C, Goebbels S, Gravel M, Nicksch E, Lee J, Braun PE, Griffiths IR, Nave KA. 2003. Disruption of Cnp1 uncouples oligodendroglial functions in axonal support and myelination. Nat Genet, 33(3), 366-374. doi:10.1038/ng1095.

Lazo OM, Gonzalez A, Ascaño M, Kuruvilla R, Couve A, Bronfman FC. 2013. BDNF regulates Rab11-mediated recycling endosome dynamics to induce dendritic branching. J Neurosci, 33(14), 6112-6122. doi:10.1523/JNEUROSCI.4630-12.2013.

Le Bras B, Barallobre MJ, Homman-Ludiye J, Ny A, Wyns S, Tammela T, Haiko P, Karkkainen MJ, Yuan L, Muriel MP, Chatzopoulou E, Bréant C, Zalc B, Carmeliet P, Alitalo K, Eichmann A, Thomas JL. 2006. VEGF-C is a trophic factor for neural progenitors in the vertebrate embryonic brain. Nat Neurosci, 9(3), 340-348. doi:10.1038/nn1646.

Le-Niculescu H, Levey DF, Ayalew M, Palmer L, Gavrin LM, Jain N, Winiger E, Bhosrekar S, Shankar G, Radel M, Bellanger E, Duckworth H, Olesek K, Vergo J, Schweitzer R, Yard M, Ballew A, Shekhar A, Sandusky GE, Schork NJ, Kurian SM, Salomon DR, Niculescu AB. 2013. Discovery and validation of blood biomarkers for suicidality. Mol Psychiatry, 18(12), 1249-1264. doi:10.1038/mp.2013.95.

Lee J, Gravel M, Zhang R, Thibault P, Braun PE. 2005. Process outgrowth in oligodendrocytes is mediated by CNP, a novel microtubule assembly myelin protein. J Cell Biol, 170(4), 661-673. doi:10.1083/jcb.200411047.

Lee J, Gray A, Yuan J, Luoh SM, Avraham H, Wood WI. 1996. Vascular endothelial growth factor-related protein: a ligand and specific activator of the tyrosine kinase receptor Flt4. Proc Natl Acad Sci U S A, 93(5), 1988-1992. doi:10.1073/pnas.1714341115.

Lefebvre L, Reader SM, Sol D. 2004. Brains, innovations and evolution in birds and primates. Brain Behav Evol, 63(4), 233-246. doi:10.1159/000076784.

Lemasters JJ, Theruvath TP, Zhong Z, Nieminen AL. 2009. Mitochondrial calcium and the permeability transition in cell death. Biochim Biophys Acta, 1787(11), 1395-1401. doi:10.1016/j.bbabio.2009.06.009.

Lewitus E, Kelava I, Kalinka AT, Tomancak P, Huttner WB. 2014. An adaptive threshold in mammalian neocortical evolution. *PLoS Biol, 12*(11). doi: 10.1371/journal.pbio.1002000.

Li W, Zhang Q, Oiso N, Novak EK, Gautam R, O'Brien EP, Tinsley CL, Blake DJ, Spritz RA, Copeland NG, Jenkins NA, Amato D, Roe BA, Starcevic M, Dell'Angelica EC, Elliott RW, Mishra V, Kingsmore SF, Paylor RE, Swank RT. 2003. Hermansky-Pudlak syndrome type 7 (HPS-7) results from mutant dysbindin, a member of the biogenesis of lysosome-related organelles complex 1 (BLOC-1). Nat Genet, 35(1), 84-89. doi:10.1038/ng1229.

Lieberman D. 2014. The Story of the Human Body: Evolution, Health, and Disease. Vintage Books, New York.

Lieberman P. 2016. The evolution of language and thought. J Anthropol Sci, 94, 127-146. doi:10.4436/jass.94029.

Linhoff MW, Laurén J, Cassidy RM, Dobie FA, Takahashi H, Nygaard HB, Airaksinen MS, Strittmatter SM, Craig AM. 2009. An Unbiased Expression Screen for Synaptogenic Proteins Identifies the LRRTM Protein Family as Synaptic Organizers. Neuron, 61(5), 734-749. doi:10.1016/j.neuron.2009.01.017.

Lo HF, Tsai CY, Chen CP, Wang LJ, Lee YS, Chen CY, Liang CT, Cheong ML, Chen H. 2017. Association of dysfunctional synapse defective 1 (SYDE1) with restricted fetal growth - SYDE1 regulates placental cell migration and invasion. J Pathol, 241(3), 324-336. doi:10.1002/path.4835.

Long M. 2001. Evolution of novel genes. Curr Opin Genet Dev, 11(6), 673-680.

Lopes LJS, Tesser-Gamba F, Petrilli AS, de Seixas Alves MT, Garcia-Filho RJ, Toledo SRC. 2017. MAPK pathways regulation by DUSP1 in the development of osteosarcoma: Potential markers and therapeutic targets. Mol Carcinog, 56(6), 1630-1641. doi:10.1002/mc.22619.

Lukaszewicz A, Savatier P, Cortay V, Giroud P, Huissoud C, Berland M, Kennedy H, Dehay C. 2005. G1 phase regulation, area-specific cell cycle control, and cytoarchitectonics in the primate cortex. Neuron, 47(3), 353-364.

Lynch M, Conery JS. 2000. The evolutionary fate and consequences of duplicate genes. Science, 290(5494), 1151-1155.

Lynch M, Katju V. 2004. The altered evolutionary trajectories of gene duplicates. Trends Genet, 20(11), 544-549. doi:10.1016/j.tig.2004.09.001.

Malatesta P, Götz M, Barsacchi G, Price J, Zoncu R, Cremisi F. 2000. PC3 overexpression affects the pattern of cell division of rat cortical precursors. Mech Dev, 90(1), 17-28.

Mansky KC, Sulzbacher S, Purdom G, Nelsen L, Hume DA, Rehli M, Ostrowski MC. 2002. The microphthalmia transcription factor and the related helix-loop-helix zipper factors TFE-3 and TFE-C collaborate to activate the tartrate-resistant acid phosphatase promoter. J Leukoc Biol, 71(2), 304-310.

Marino L. 1998. A comparison of encephalization between odontocete cetaceans and anthropoid primates. Brain Behav Evol, 51(4), 230-238.

Marino L. 2002. Convergence of complex cognitive abilities in cetaceans and primates. Brain Behav Evol, 59(1-2), 21-32. doi:10.1159/000063731.

Marques-Bonet T, Kidd JM, Ventura M, Graves TA, Cheng Z, Hillier LW, Jiang Z, Baker C, Malfavon-Borja R, Fulton LA, Alkan C, Aksay G, Girirajan S, Siswara P, Chen L, Cardone MF, Navarro A, Mardis ER, Wilson RK, Eichler EE. 2009. A burst of segmental duplications in the genome of the African great ape ancestor. Nature, 457(7231), 877-881. doi:10.1038/nature07744.

Marthiens V, ffrench-Constant C. 2009. Adherens junction domains are split by asymmetric division of embryonic neural stem cells. EMBO Rep., 10(5), 515-520. doi:10.1038/embor.2009.36.

Martinez A, Yamashita S, Nagaike T, Sakaguchi Y, Suzuki T, Tomita K. 2017. Human BCDIN3D monomethylates cytoplasmic histidine transfer RNA. Nucleic Acids Res, 45(9), 5423-5436. doi:10.1093/nar/gkx051.

Martynoga B, Drechsel D, Guillemot F. 2012. Molecular control of neurogenesis: a view from the mammalian cerebral cortex. Cold Spring Harb Perspect Biol, 4(10). doi:10.1101/cshperspect.a008359.

McLysaght A, Hokamp K, Wolfe KH. 2002. Extensive genomic duplication during early chordate evolution. Nat Genet, 31(2), 200-204. doi:10.1038/ng884.

Menon SG, Sarsour EH, Kalen AL, Venkataraman S, Hitchler MJ, Domann FE, Oberley LW, Goswami PC. 2007. Superoxide signaling mediates N-acetyl-L-cysteine-induced G1 arrest: regulatory role of cyclin D1 and manganese superoxide dismutase. Cancer Res, 67(13), 6392-6399. doi:10.1158/0008-5472.Can-07-0225.

Meredith RW, Janecka JE, Gatesy J, Ryder OA, Fisher CA, Teeling EC, Goodbla A, Eizirik E, Simao TL, Stadler T, Rabosky DL, Honeycutt RL, Flynn JJ, Ingram CM, Steiner C, Williams TL, Robinson TJ, Burk-Herrick A, Westerman M, Ayoub NA, Springer MS, Murphy WJ. 2011. Impacts of the Cretaceous Terrestrial Revolution and KPg extinction on mammal diversification. Science, 334(6055), 521-524. doi:10.1126/science.1211028.

Mitkus SN, Hyde TM, Vakkalanka R, Kolachana B, Weinberger DR, Kleinman JE, Lipska BK. 2008. Expression of oligodendrocyte-associated genes in dorsolateral prefrontal cortex of patients with schizophrenia. Schizophr Res, 98(1), 129-138. doi:10.1016/j.schres.2007.09.032.

Momburg F, Tan P. 2002. Tapasin-the keystone of the loading complex optimizing peptide binding by MHC class I molecules in the endoplasmic reticulum. Mol Immunol, 39(3-4), 217-233.

Monoh K, Kurihara T, Takahashi Y, Ichikawa T, Kumanishi T, Hayashi S, Minoshima S, Shimizu N. 1993. Structure, expression and chromosomal localization of the gene encoding human 2',3'-cyclic-nucleotide 3'-phosphodiesterase. Gene, 129(2), 297-301. doi:10.1016/0378-1119(93)90283-9.

Montiel JF, Vasistha NA, Garcia-Moreno F, Molnar Z. 2016. From sauropsids to mammals and back: New approaches to comparative cortical development. J Comp Neurol, 524(3), 630-645. doi:10.1002/cne.23871.

Morante-Redolat JM, Fariñas I. 2016. Fetal neurogenesis: breathe HIF you can. EMBO J, 35(9), 901-903. doi:10.15252/embj.201694238.

Morita M, Prudent J, Basu K, Goyon V, Katsumura S, Hulea L, Pearl D, Siddiqui N, Strack S, McGuirk S, St-Pierre J, Larsson O, Topisirovic I, Vali H, McBride HM, Bergeron JJ, Sonenberg

N. 2017. mTOR Controls Mitochondrial Dynamics and Cell Survival via MTFP1. Mol Cell, 67(6), 922-935.e925. doi:10.1016/j.molcel.2017.08.013.

Mortensen HS, Pakkenberg B, Dam M, Dietz R, Sonne C, Mikkelsen B, Eriksen N. 2014. Quantitative relationships in delphinid neocortex. Front Neuroanat, 8(46), 132. doi:10.3389/fnana.2014.00132.

Myllykoski M, Seidel L, Muruganandam G, Raasakka A, Torda AE, Kursula P. 2016. Structural and functional evolution of 2',3'-cyclic nucleotide 3'-phosphodiesterase. Brain Res, 1641, 64-78. doi:10.1016/j.brainres.2015.09.004.

Namba T, Dóczi J, Pinson A, Xing L, Kalebic N, Wilsch-Braeuninger M, Long KR, Vaid S, Lauer J, Bogdanova A, Borgonovo B, Shevchenko A, Keller P, Drechsel D, Kurzchalia T, Wimberger P, Chinopoulos C,. Huttner WB. Human-specific ARHGAP11B acts in mitochondria to expand neocortical progenitors by glutaminolysis. 2019. submitted.

Namba T, Huttner WB. 2017. Neural progenitor cells and their role in the development and evolutionary expansion of the neocortex. Wiley Interdiscip Rev Dev Biol, 6(1), e256. doi:10.1002/wdev.256.

Navarrete A, van Schaik CP, Isler K. 2011. Energetics and the evolution of human brain size. Nature, 480(7375), 91-93. doi:10.1038/nature10629.

Necsulea A, Soumillon M, Warnefors M, Liechti A, Daish T, Zeller U, Baker JC, Grutzner F, Kaessmann H. 2014. The evolution of lncRNA repertoires and expression patterns in tetrapods. Nature, 505(7485), 635-640. doi:10.1038/nature12943.

Neefjes J, Jongsma MLM, Paul P, Bakke O. 2011. Towards a systems understanding of MHC class I and MHC class II antigen presentation. Nat Rev Immunol, 11(12), 823-836. doi:10.1038/nri3084.

Nehar S, Mishra M, Heese K. 2009. Identification and characterisation of the novel amyloid-beta peptide-induced protein p17. FEBS Lett, 583(19), 3247-3253. doi:10.1016/j.febslet.2009.09.018.

Nei M, Xu P, Glazko G. 2001. Estimation of divergence times from multiprotein sequences for a few mammalian species and several distantly related organisms. Proc Natl Acad Sci U S A, 98(5), 2497-2502. doi:10.1073/pnas.051611498.

Nei M, Zhang J, Yokoyama S. 1997. Color vision of ancestral organisms of higher primates. Mol Biol Evol, 14(6), 611-618. doi:10.1093/oxfordjournals.molbev.a025800.

Newell EA, Exo JL, Verrier JD, Jackson TC, Gillespie DG, Janesko-Feldman K, Kochanek P M, Jackson EK. 2015. 2',3'-cAMP, 3'-AMP, 2'-AMP and adenosine inhibit TNF-α and CXCL10 production from activated primary murine microglia via A2A receptors. Brain Res, 1594, 27-35. doi:10.1016/j.brainres.2014.10.059.

Nguyen AL, Gentilello AS, Balboula AZ, Shrivastava V, Ohring J, Schindler K. 2014. Phosphorylation of threonine 3 on histone H3 by haspin kinase is required for meiosis I in mouse oocytes. J Cell Sci, 127(Pt 23), 5066-5078. doi:10.1242/jcs.158840.

Niculescu AB, Le-Niculescu H, Levey DF, Phalen PL, Dainton HL, Roseberry K, Niculescu EM, Niezer JO, Williams A, Graham DL, Jones TJ, Venugopal V, Balle A, Yard M, Gelbart T, Kurian SM, Shekhar A, Schork NJ, Sandusky GE, Salomon DR. 2017. Precision medicine for suicidality: from universality to subtypes and personalization. Mol Psychiatry, 22(9), 1250-1273. doi:10.1038/mp.2017.128.

Noback CR, Strominger NL, Demarest RJ, Ruggiero DA. 2005. The Human Nervous System: Structure and Function (Sixth ed.). Humana Press, Totowa (New Jersey).

Noctor SC, Martinez-Cerdeno V, Ivic L, Kriegstein AR. 2004. Cortical neurons arise in symmetric and asymmetric division zones and migrate through specific phases. Nat Neurosci, 7(2), 136-144.

Nonaka-Kinoshita M, Reillo I, Artegiani B, Martinez-Martinez MA, Nelson M, Borrell V, Calegari F. 2013. Regulation of cerebral cortex size and folding by expansion of basal progenitors. EMBO J, 32(13), 1817-1828. doi:10.1038/emboj.2013.96.

O'Brien KM, Cole SR, Engel LS, Bensen JT, Poole C, Herring AH, Millikan RC. 2014. Breast cancer subtypes and previously established genetic risk factors: a bayesian approach. Cancer Epidemiol Biomarkers Prev, 23(1), 84-97. doi:10.1158/1055-9965.Epi-13-0463.

O'Leary MA, Bloch JI, Flynn JJ, Gaudin TJ, Giallombardo A, Giannini NP, Goldberg SL, Kraatz BP, Luo ZX, Meng J, Ni X, Novacek MJ, Perini FA, Randall ZS, Rougier GW, Sargis EJ, Silcox MT, Simmons NB, Spaulding M, Velazco PM, Weksler M, Wible JR, Cirranello AL. 2013. The placental mammal ancestor and the post-K-Pg radiation of placentals. Science, 339(6120), 662-667. doi:10.1126/science.1229237.

Oettinghaus B, Licci M, Scorrano L, Frank S. 2012. Less than perfect divorces: dysregulated mitochondrial fission and neurodegeneration. Acta Neuropathol, *123*(2), 189-203. doi:10.1007/s00401-011-0930-z.

Ohno S. (1967). Sex Chromosomes and Sex-Linked Genes. Springer, Berlin.

Ohno S. (1970). Evolution by gene duplication. Springer, Berlin.

Olafson RW, Drummond GI, Lee JF. 2011. Studies on 2',3'-cyclic nucleotide-3'-phosphohydrolase from brain. Can J Biochem, 47(10), 961-966. doi:10.1139/o69-151.

Ortmann B, Copeman J, Lehner PJ, Sadasivan B, Herberg JA, Grandea AG, Riddell SR, Tampé R, Spies T, Trowsdale J, Cresswell P. 1997. A critical role for tapasin in the assembly and function of multimeric MHC class I-TAP complexes. Science, 277(5330), 1306-1309.

Osumi N, Shinohara H, Numayama-Tsuruta K, Maekawa M. 2008. Concise review: Pax6 transcription factor contributes to both embryonic and adult neurogenesis as a multifunctional regulator. Stem Cells, *26*(7), 1663-1672. doi:10.1634/stemcells.2007-0884.

Palmer TD, Willhoite AR, Gage FH. 2000. Vascular niche for adult hippocampal neurogenesis. J Comp Neurol, 425(4), 479-494.

Papp B, Pál C, Hurst LD. 2003. Dosage sensitivity and the evolution of gene families in yeast. Nature, 424(6945), 194-197. doi:10.1038/nature01771.

Paridaen JTML, Wilsch-Bräuninger M, Huttner WB. 2013. Asymmetric Inheritance of Centrosome-Associated Primary Cilium Membrane Directs Ciliogenesis after Cell Division. Cell, *155*(2), 333-344. doi:10.1016/j.cell.2013.08.060.

Paterson A, Chapman B, Kissinger J, Bowers J, Feltus F, Estill J. 2006. Many gene and domain families have convergent fates following independent whole-genome duplication events in Arabidopsis, Oryza, Saccharomyces and Tetraodon. Trends Genet, 22(11), 597-602. doi:10.1016/j.tig.2006.09.003.

Peirce TR, Bray NJ, Williams NM, Norton N, Moskvina V, Preece A, Haroutunian V, Buxbaum JD, Owen MJ, O'Donovan M. C. 2006. Convergent evidence for 2',3'-Cyclic Nucleotide 3'-phosphodiesterase as a possible susceptibility gene for schizophrenia. Arch Gen Psychiatry, 63(1), 18-24. doi:10.1001/archpsyc.63.1.18.

Pilaz LJ, Patti D, Marcy G, Ollier E, Pfister S, Douglas RJ, Betizeau M, Gautier E, Cortay V, Doerflinger N, Kennedy H, Dehay C. 2009. Forced G1-phase reduction alters mode of division, neuron number, and laminar phenotype in the cerebral cortex. Proc. Natl. Acad. Sci. USA, *106*(51), 21924-21929.

Pilz GA, Shitamukai A, Reillo I, Pacary E, Schwausch J, Stahl R, Ninkovic J, Snippert HJ, Clevers H, Godinho L, Guillemot F, Borrell V, Matsuzaki F, Götz M. 2013. Amplification of progenitors in the mammalian telencephalon includes a new radial glial cell type. Nat Commun, 4, 2125. doi:10.1038/ncomms3125.

Pollen AA, Nowakowski TJ, Chen J, Retallack H, Sandoval-Espinosa C, Nicholas CR, Shuga J, Liu SJ, Oldham MC, Diaz A, Lim DA, Leyrat AA, West JA, Kriegstein AR. 2015. Molecular Identity of Human Outer Radial Glia during Cortical Development. Cell, 163(1), 55-67. doi:10.1016/j.cell.2015.09.004.

Pratt ZL, Zhang J, Sugden B. 2012. The latent membrane protein 1 (LMP1) oncogene of Epstein-Barr virus can simultaneously induce and inhibit apoptosis in B cells. J Virol, 86(8), 4380-4393. doi:10.1128/JVI.06966-11.

Proenca CC, Gao KP, Shmelkov SV, Rafii S, Lee FS. 2011. Slitrks as emerging candidate genes involved in neuropsychiatric disorders. Trends Neurosci, 34(3), 143-153. doi:10.1016/j.tins.2011.01.001.

Puehringer D, Orel N, Lüningschrör P, Subramanian N, Herrmann T, Chao MV, Sendtner M. 2013. EGF transactivation of Trk receptors regulates the migration of newborn cortical neurons. Nat Neurosci, 16(4), 407-415. doi:10.1038/nn.3333.

Qian W, Zhang J. 2014. Genomic evidence for adaptation by gene duplication. Genome Res, 24(8), 1356-1362. doi:10.1101/gr.172098.114.

Qu JJ, Qu XY, Zhou DZ. 2017. miR-4262 inhibits colon cancer cell proliferation via targeting of GALNT4. Mol Med Report, 16(4), 3731-3736. doi:10.3892/mmr.2017.7057.

Quartuccio SM, Dipali SS, Schindler K. 2017. Haspin inhibition reveals functional differences of interchromatid axis-localized AURKB and AURKC. Mol Biol Cell, 28(17), 2233-2240. doi:10.1091/mbc.E16-12-0850.

Williams RW, Herrup K. 1988. The Control of Neuron Number. Annu Rev Neurosci, 11(1), 423-453. doi:10.1146/annurev.ne.11.030188.002231.

Raab S, Plate KH. 2007. Different networks, common growth factors: shared growth factors and receptors of the vascular and the nervous system. Acta Neuropathol, 113(6), 607-626. doi:10.1007/s00401-007-0228-3.

Rajkowska G, Mahajan G, Maciag D, Sathyanesan M, Iyo AH, Moulana M, Kyle PB, Woolverton WL, Miguel-Hidalgo JJ, Stockmeier CA, Newton SS. 2015. Oligodendrocyte morphometry and expression of myelin - Related mRNA in ventral prefrontal white matter in major depressive disorder. J Psychiatr Res, 65, 53-62. doi:10.1016/j.jpsychires.2015.04.010.

Rakic P. 1972. Mode of cell migration to the superficial layers of fetal monkey neocortex. J Comp Neurol, 145(1), 61-83.

Rakic P. 1995. A small step for the cell, a giant leap for mankind: a hypothesis of neocortical expansion during evolution. Trends Neurosci, 18, 383-388.

Rakic P. 2009. Evolution of the neocortex: a perspective from developmental biology. Nat Rev Neurosci, 10(10), 724-735. doi:10.1038/nrn2719.

Rasola A, Bernardi P. 2011. Mitochondrial permeability transition in Ca^{2+} - dependent apoptosis and necrosis. Cell Calcium, 50(3), 222-233. doi:10.1016/j.ceca.2011.04.007.

Rasola A, Bernardi P. 2014. The mitochondrial permeability transition pore and its adaptive responses in tumor cells. Cell Calcium, 56(6), 437-445. doi:10.1016/j.ceca.2014.10.003.

Reader SM, Laland KN. 2002. Social intelligence, innovation, and enhanced brain size in primates. Proc Natl Acad Sci U S A, 99(7), 4436-4441.

Reillo I, De Juan Romero C, García-Cabezas MÁ, Borrell V. 2011. A role for intermediate radial glia in the tangential expansion of the mammalian cerebral cortex. *Cerebral cortex, 21*(7), 1674-1694. doi:10.1093/cercor/bhq238.

Reyes LD, Sherwood CC. 2015. Neuroscience and Human Brain Evolution. In E. Bruner (Ed.), Human Paleoneurology (pp. 11-37). Springer International Publishing, Cham.

Rincic M, Rados M, Krsnik Z, Gotovac K, Borovecki F, Liehr T, Brecevic L. 2016. Complex intrachromosomal rearrangement in 1q leading to 1q32.2 microdeletion: a potential role of SRGAP2 in the gyrification of cerebral cortex. Mol Cytogenet, 9(1), 19. doi:10.1186/s13039-016-0221-4.

Robert I, Gaudot L, Yélamos J, Noll A, Wong HK, Dantzer F, Schreiber V, Reina San Martin B. (2017). Robust immunoglobulin class switch recombination and end joining in Parp9-deficient mice. Eur J Immunol, 47(4), 665-676. doi:10.1002/eji.201646757.

Roth G, Dicke U. 2005. Evolution of the brain and intelligence. Trends Cogn Sci, 9(5), 250-257. doi:10.1016/j.tics.2005.03.005.

Rouquier S, Blancher A, Giorgi D. 2000. The olfactory receptor gene repertoire in primates and mouse: evidence for reduction of the functional fraction in primates. Proc Natl Acad Sci U S A, 97(6), 2870-2874. doi:10.1073/pnas.040580197.

Santos-Rosa H, Schneider R, Bannister AJ, Sherriff J, Bernstein BE, Emre NC, Schreiber SL, Mellor J, Kouzarides T. 2002. Active genes are tri-methylated at K4 of histone H3. Nature, 419(6905), 407-411. doi:10.1038/nature01080.

Sarsour EH, Kumar MG, Chaudhuri L, Kalen AL, Goswami PC. 2009. Redox control of the cell cycle in health and disease. Antioxid Redox Signal, 11(12), 2985-3011. doi:10.1089/ars.2009.2513.

Schmitz SU, Albert M, Malatesta M, Morey L, Johansen JV, Bak M, Tommerup N, Abarrategui , Helin K. 2011. Jarid1b targets genes regulating development and is involved in neural differentiation. EMBO J, 30(22), 4586-4600. doi:10.1038/emboj.2011.383.

Seizl M, Larivière L, Pfaffeneder T, Wenzeck L, Cramer P. 2011. Mediator head subcomplex Med11/22 contains a common helix bundle building block with a specific function in transcription initiation complex stabilization. Nucleic Acids Res, 39(14), 6291-6304. doi:10.1093/nar/gkr229.

Shen J, Zhou S, Shi L, Liu X, Lin H, Yu H, Xiaoliang, Tang J, Yu T, Cai X. 2017. DUSP1 inhibits cell proliferation, metastasis and invasion and angiogenesis in gallbladder cancer. Oncotarget, 8(7), 12133-12144. doi:10.18632/oncotarget.14815.

Shen Q, Goderie SK, Jin L, Karanth N, Sun Y, Abramova N, Vincent P, Pumiglia K, Temple S. 2004. Endothelial cells stimulate self-renewal and expand neurogenesis of neural stem cells. Science, 304(5675), 1338-1340.

Shin YJ, Choi JS, Choi JY, Hou Y, Cha JH, Chun MH, Lee MY. 2010. Induction of vascular endothelial growth factor receptor-3 mRNA in glial cells following focal cerebral ischemia in rats. J Neuroimmunol, 229(1-2), 81-90. doi:10.1016/j.jneuroim.2010.07.008.

Shin YJ, Choi JS, Lee JY, Choi JY, Cha JH, Chun MH, Lee MY. 2008. Differential regulation of vascular endothelial growth factor-C and its receptor in the rat hippocampus following transient forebrain ischemia. Acta Neuropathol, 116(5), 517-527. doi:10.1007/s00401-008-0423-x.

Shmelkov SV, Hormigo A, Jing D, Proenca CC, Bath KG, Milde T, Shmelkov E, Kushner JS, Baljevic M, Dincheva I, Murphy AJ, Valenzuela DM, Gale NW, Yancopoulos GD, Ninan I, Lee FS, Rafii S. 2010. Slitrk5 deficiency impairs corticostriatal circuitry and leads to obsessive-compulsive-like behaviors in mice. Nat Med, 16(5), 598-602- 591p following 602. doi:10.1038/nm.2125.

Sol D. 2009. Revisiting the cognitive buffer hypothesis for the evolution of large brains. Biol Lett, 5(1), 130-133. doi:10.1098/rsbl.2008.0621.

Sondell M, Lundborg G, Kanje M. 1999. Vascular endothelial growth factor has neurotrophic activity and stimulates axonal outgrowth, enhancing cell survival and Schwann cell proliferation in the peripheral nervous system. J Neurosci, 19(14), 5731-5740.

Song M, Giza J, Proenca CC, Jing D, Elliott M, Dincheva I, Shmelkov SV, Kim J, Schreiner R, Huang SH, Castrén E, Prekeris R, Hempstead BL, Chao MV, Dictenberg JB, Rafii S, Chen ZY, Rodriguez-Boulan E, Lee FS. 2015. Slitrk5 Mediates BDNF-Dependent TrkB Receptor Trafficking and Signaling. Dev Cell, 33(6), 690-702. doi:10.1016/j.devcel.2015.04.009.

Song M, Mathews CA, Stewart SE, Shmelkov SV, Mezey JG, Rodriguez-Flores JL, Rasmussen SA, Britton JC, Oh YS, Walkup JT, Lee FS, Glatt CE. 2017. Rare Synaptogenesis-Impairing Mutations in SLITRK5 Are Associated with Obsessive Compulsive Disorder. PLoS One, 12(1), e0169994. doi:10.1371/journal.pone.0169994.

Song W, Chen J, Petrilli A, Liot G, Klinglmayr E, Zhou Y, Poquiz P, Tjong J, Pouladi MA, Hayden MR, Masliah E, Ellisman M, Rouiller I, Schwarzenbacher R, Bossy B, Perkins G,

Bossy-Wetzel E. 2011. Mutant huntingtin binds the mitochondrial fission GTPase dynamin-related protein-1 and increases its enzymatic activity. Nat Med, 17(3), 377-382. doi:10.1038/nm.2313.

Sousa AMM, Meyer KA, Santpere G, Gulden FO, Šestan N. 2017. Evolution of the Human Nervous System Function, Structure, and Development. Cell, 170(2), 226-247. doi:10.1016/j.cell.2017.06.036.

Stahl R, Walcher T, De Juan Romero C, Pilz GA, Cappello S, Irmler M, Sanz-Aquela JM, Beckers J, Blum R, Borrell V, Götz M. 2013. Trnp1 regulates expansion and folding of the Mammalian cerebral cortex by control of radial glial fate. Cell, 153(3), 535-549. doi:10.1016/j.cell.2013.03.027.

Stedman HH. Kozyak BW, Nelson A, Thesier DM, Su LT, Low DW, Bridges CR, Shrager JB, Minugh-Purvis N, Mitchell MA. 2004. Myosin gene mutation correlates with anatomical changes in the human lineage. Nature, 428(6981), 415-418. doi:10.1038/nature02358.

Steib K, Schaffner I, Jagasia R, Ebert B, Lie DC. 2014. Mitochondria modify exercise-induced development of stem cell-derived neurons in the adult brain. J Neurosci, 34(19), 6624-6633. doi:10.1523/jneurosci.4972-13.2014.

Steingrimsson E, Tessarollo L, Pathak B, Hou L, Arnheiter H, Copeland NG, Jenkins NA. 2002. Mitf and Tfe3, two members of the Mitf-Tfe family of bHLH-Zip transcription factors, have important but functionally redundant roles in osteoclast development. Proc Natl Acad Sci U S A, 99(7), 4477-4482. doi:10.1073/pnas.072071099.

Stimpson NJ, Davison G, Javadi AH. 2018. Joggin' the Noggin: Towards a Physiological Understanding of Exercise-Induced Cognitive Benefits. Neurosci Biobehav Rev, 88, 177-186. doi:10.1016/j.neubiorev.2018.03.018.

Stone TW, Forrest CM, Mackay GM, Stoy N, Darlington LG. 2007. Tryptophan, adenosine, neurodegeneration and neuroprotection. Metab Brain Dis, 22(3-4), 337-352. doi:10.1007/s11011-007-9064-3.

Striedter GF. 2005. Principles of brain evolution. Sinauer, Sunderland (Mass.).

Subramanian J, Nedivi E. 2016. Filling the (SR)GAP in Excitatory/Inhibitory Balance. Neuron, 91(2), 205-207. doi:10.1016/j.neuron.2016.07.008.

Sudmant PH, Huddleston J, Catacchio CR, Malig M, Hillier LW, Baker C, Mohajeri K, Kondova I, Bontrop RE, Persengiev S, Antonacci F, Ventura M, Prado-Martinez J, Project GAG, Marques-Bonet T, Eichler EE. 2013. Evolution and diversity of copy number variation in the great ape lineage. Genome Res, 23(9), 1373-1382. doi:10.1101/gr.158543.113.

Sudmant PH, Kitzman JO, Antonacci F, Alkan C, Malig M, Tsalenko A, Sampas N, Bruhn L, Shendure J, Project G, Eichler EE. 2010. Diversity of human copy number variation and multicopy genes. Science, 330(6004), 641-646. doi:10.1126/science.1197005.

Suh H, Consiglio A, Ray J, Sawai T, D'Amour K A, Gage FH. (2007). In Vivo Fate Analysis Reveals the Multipotent and Self-Renewal Capacities of Sox2+ Neural Stem Cells in the Adult Hippocampus. Cell Stem Cell, *1*(5), 515-528. doi:10.1016/j.stem.2007.09.002.

Sun T, Hevner RF. 2014. Growth and folding of the mammalian cerebral cortex: from molecules to malformations. Nat Rev Neurosci, 15(4), 217-232. doi:10.1038/nrn3707.

Sun Y, Jin K, Xie L, Childs J, Mao XO, Logvinova A, Greenberg DA. 2003. VEGF-induced neuroprotection, neurogenesis, and angiogenesis after focal cerebral ischemia. J Clin Invest, 111(12), 1843-1851. doi:10.1172/JCI17977.

Sundaresan M, Yu ZX, Ferrans VJ, Irani K, Finkel T. 1995. Requirement for generation of H_2O_2 for platelet-derived growth factor signal transduction. Science, 270(5234), 296-299.

Szmulewicz MN, Novick GE, Herrera RJ. 1998. Effects of Alu insertions on gene function. Electrophoresis, 19(8-9), 1260-1264. doi:10.1002/elps.1150190806.

Takahashi H, Craig AM. 2013. Protein tyrosine phosphatases PTPδ, PTPσ, and LAR: presynaptic hubs for synapse organization. Trends Neurosci, 36(9), 522-534. doi:10.1016/j.tins.2013.06.002.

Takahashi H, Katayama KI, Sohya K, Miyamoto H, Prasad T, Matsumoto Y, Ota M, Yasuda H, Tsumoto T, Aruga J, Craig AM. 2012. Selective control of inhibitory synapse development by Slitrk3-PTPδ trans-synaptic interaction. Nat Neurosci, 15(3), 389-398- S381-382. doi:10.1038/nn.3040.

Takahashi T, Nowakowski RS, Caviness VS Jr. 1995. The cell cycle of the pseudostratified ventricular epithelium of the embryonic murine cerebral wall. J Neurosci, 15(9), 6046-6057.

Tang Y, Wang J, Zhang Y, Zhuo M, Song L, Tang Z, Zang G, Chen X, Yu Y. 2016. Correlation between low tapasin expression and impaired CD8+ T-cell function in patients with chronic hepatitis B. Mol Med Report, 14(4), 3315-3322. doi:10.3892/mmr.2016.5610.

Terrinoni A, Ranalli M, Cadot B, Leta A, Bagetta G, Vousden KH, Melino G. 2004. p73-alpha is capable of inducing scotin and ER stress. *Oncogene, 23(20)*, 3721-3725. doi:10.1038/sj.onc.1207342.

Thomaidou D, Mione MC, Cavanagh JFR, Parnavelas JG. 1997. Apoptosis and Its Relation to the Cell Cycle in the Developing Cerebral Cortex. J Neurosci, 17(3), 1075-1085. doi:10.1016/0092-8674(92)90531-G.

Tkachev D, Mimmack ML, Ryan MM, Wayland M, Freeman T, Jones PB, Starkey M, Webster MJ, Yolken RH, Bahn S. 2003. Oligodendrocyte dysfunction in schizophrenia and bipolar disorder. The Lancet, 362(9386), 798-805. doi:10.1016/S0140-6736(03)14289-4.

Toma JS, McPhail LT, Ramer MS. 2007. Differential RIP antigen (CNPase) expression in peripheral ensheathing glia. Brain Res, 1137, 1-10. doi:10.1016/j.brainres.2006.12.053.

Tondera D, Czauderna F, Paulick K, Schwarzer R, Kaufmann J, Santel A. 2005. The mitochondrial protein MTP18 contributes to mitochondrial fission in mammalian cells. J Cell Sci, 118(Pt 14), 3049-3059. doi:10.1242/jcs.02415.

Tondera D, Santel A, Schwarzer R, Dames S, Giese K, Klippel A, Kaufmann J. 2004. Knockdown of MTP18, a novel phosphatidylinositol 3-kinase-dependent protein, affects mitochondrial morphology and induces apoptosis. J Biol Chem, 279(30), 31544-31555. doi:10.1074/jbc.M404704200.

Tong L, Balazs R, Thornton PL, Cotman CW. 2004. β-Amyloid Peptide at Sublethal Concentrations Downregulates Brain-Derived Neurotrophic Factor Functions in Cultured Cortical Neurons. J Neurosci, 24(30), 6799-6809. doi:10.1523/JNEUROSCI.5463-03.2004.

Torrents D, Suyama M, Zdobnov E, Bork P. 2003. A genome-wide survey of human pseudogenes. Genome Res, 13(12), 2559-2567. doi:10.1101/gr.1455503.

Vaid S, Camp JG, Hersemann L, Eugster C, Heninger AK, Winkler S, Brandl H, Sarov M, Treutlein B, Huttner WB, Namba T. 2018. A novel population of Hopx-dependent basal radial glial cells in the developing mouse neocortex. Development, 145(20). doi:10.1242/dev.169276.

Vallender EJ. 2012. Genetic correlates of the evolving primate brain. Prog Brain Res, 195, 27-44. doi:10.1016/B978-0-444-53860-4.00002-7.

Van de Peer Y, Taylor JS, Meyer A. 2003. Are all fishes ancient polyploids? J Struct Funct Genomics, 3(1-4), 65-73.

Vasudevan A, Long JE, Crandall JE, Rubenstein JL, Bhide PG. 2008. Compartment-specific transcription factors orchestrate angiogenesis gradients in the embryonic brain. Nat Neurosci, 11(4), 429-439.

Verrier JD, Jackson TC, Gillespie DG, Janesko-Feldman K, Bansal R, Goebbels S, Nave KA, Kochanek PM, Jackson EK. 2013. Role of CNPase in the oligodendrocytic extracellular 2',3'-cAMP-adenosine pathway. Glia, 61(10), 1595-1606. doi:10.1002/glia.22523.

Wai T, Langer T. 2016. Mitochondrial Dynamics and Metabolic Regulation. Trends Endocrinol Metab, 27(2), 105-117. doi:10.1016/j.tem.2015.12.001.

Wang CH, Su PT, Du XY, Kuo MW, Lin CY, Yang CC, Chan HS, Chang SJ, Kuo C, Seo K, Leung LL, Chuang YJ. 2010. Thrombospondin type I domain containing 7A (THSD7A) mediates endothelial cell migration and tube formation. J Cell Physiol, 222(3), 685-694. doi:10.1002/jcp.21990.

Wang LT, Lin CS, Chai CY, Liu KY, Chen JY, Hsu SH. 2011. Functional interaction of Ugene and EBV infection mediates tumorigenic effects. Oncogene, 30(26), 2921-2932. doi:10.1038/onc.2011.16.

Wang X, Su B, Lee HG, Li X, Perry G, Smith MA, Zhu X. 2009. Impaired balance of mitochondrial fission and fusion in Alzheimer's disease. J Neurosci, 29(28), 9090-9103. doi:10.1523/jneurosci.1357-09.2009.

Wang X, Yan MH, Fujioka H, Liu J, Wilson-Delfosse A, Chen SG, Perry G, Casadesus G, Zhu X. 2012. LRRK2 regulates mitochondrial dynamics and function through direct interaction with DLP1. Hum Mol Genet, 21(9), 1931-1944. doi:10.1093/hmg/dds003.

Ward NL, Lamanna JC. 2004. The neurovascular unit and its growth factors: coordinated response in the vascular and nervous systems. Neurol Res, 26(8), 870-883. doi:10.1179/016164104x3798.

Warren N, Caric D, Pratt T, Clausen JA, Asavaritikrai P, Mason JO, Hill RE, Price DJ. 1999. The transcription factor, Pax6, is required for cell proliferation and differentiation in the developing cerebral cortex. Cereb Cortex, 9(6), 627-635.

Wentzel C, Sommer JE, Nair R, Stiefvater A, Sibarita JB, Scheiffele P. 2013. mSYD1A, a mammalian synapse-defective-1 protein, regulates synaptogenic signaling and vesicle docking. Neuron, 78(6), 1012-1023. doi:10.1016/j.neuron.2013.05.010.

White SH. 1992. Amino acid preferences of small proteins. Implications for protein stability and evolution. J Mol Biol, 227(4), 991-995.

Whitfeld PR, Heppel LA, Markham R. 1955. The enzymic hydrolysis of ribonucleoside-2′:3′ phosphates. Biochem J, 60(1), 15.

Xhemalce B, Robson SC, Kouzarides T. 2012. Human RNA methyltransferase BCDIN3D regulates microRNA processing. Cell, 151(2), 278-288. doi:10.1016/j.cell.2012.08.041.

Yang CS, Jividen K, Spencer A, Dworak N, Ni L, Oostdyk LT, Chatterjee M, Kuśmider B, Reon B, Parlak M, Gorbunova V, Abbas T, Jeffery E, Sherman NE, Paschal BM. 2017. Ubiquitin modification by the E3 Ligase/ADP-Ribosyltransferase Dtx3L/Parp9. Mol Cell, 66(4), 503-516.e505. doi:10.1016/j.molcel.2017.04.028.

Yang L, Kan EM, Lu J, Wu C, Ling EA. 2014. Expression of 2′,3′-cyclic nucleotide 3′-phosphodiesterase (CNPase) and its roles in activated microglia in vivo and in vitro. J Neuroinflammation, 11(1), 148. doi:10.1186/s12974-014-0148-9.

Yang P, Cai L, Zhang G, Bian Z, Han G. 2016. The role of the miR-17–92 cluster in neurogenesis and angiogenesis in the central nervous system of adults. J Neurosci Res, 8(9), 963. doi:10.1002/jnr.23991.

Yao L, Chi Y, Hu X, Li S, Qiao F, Wu J, Shao ZM. 2016. Elevated expression of RNA methyltransferase BCDIN3D predicts poor prognosis in breast cancer. Oncotarget, 7(33), 53895-53902. doi:10.18632/oncotarget.9656.

Yoon Y, Krueger EW, Oswald BJ, McNiven MA. 2003. The mitochondrial protein hFis1 regulates mitochondrial fission in mammalian cells through an interaction with the dynamin-like protein DLP1. Mol Cell Biol, 23(15), 5409-5420. doi:10.1128/MCB.23.15.5409-5420.2003.

Young JM, Friedman C, Williams EM, Ross JA, Tonnes-Priddy L, Trask BJ. 2002. Different evolutionary processes shaped the mouse and human olfactory receptor gene families. Hum Mol Genet, 11(5), 535-546. doi:10.1093/hmg/11.5.535.

Yu W. 1994. Embryonic expression of myelin genes: Evidence for a focal source of oligodendrocyte precursors in the ventricular zone of the neural tube. Neuron, 12(6), 1353-1362. doi:10.1016/0896-6273(94)90450-2.

Zhang J. 2003. Evolution by gene duplication: an update. Trends Ecol Evol, 18(6), 292-298. doi:10.1016/S0169-5347(03)00033-8.

Zhang J, Zhang Z, Wang Q, Xing XJ, Zhao Y. 2016. Overexpression of microRNA-365 inhibits breast cancer cell growth and chemo-resistance through GALNT4. Eur Rev Med Pharmacol Sci, 20(22), 4710-4718.

Zhang P, Gu Z, Li WH. 2003. Different evolutionary patterns between young duplicate genes in the human genome. Genome Biol, 4(9), R56. doi:10.1186/gb-2003-4-9-r56.

Zhang Y, Mao D, Roswit WT, Jin X, Patel AC, Patel DA, Agapov E, Wang Z, Tidwell RM, Atkinson JJ, Huang G, McCarthy R, Yu J, Yun NE, Paessler S, Lawson TG, Omattage NS, Brett TJ, Holtzman MJ. 2015. PARP9-DTX3L ubiquitin ligase targets host histone H2BJ and viral 3C protease to enhance interferon signaling and control viral infection. Nat Immunol, 16(12), 1215-1227. doi:10.1038/ni.3279.

Zhang Y, Williams DB. 2006. Assembly of MHC class I molecules within the endoplasmic reticulum. Immunol Res, 35(1-2), 151-162. doi:10.1385/IR:35:1:151.

Zocchi L, Bourdon JC, Codispoti A, Knight R, Lane DP, Melino G, Terrinoni A. 2008. Scotin: A new p63 target gene expressed during epidermal differentiation. Biochem Biophys Res Commun, 367(2), 271-276. doi:10.1016/j.bbrc.2007.12.115.

11 Appendix

11.1 Conference presentation

2017 November 23 – 24
14th International Medical Postgraduate Conference (*New Frontiers in the Reseach of Ph.D. Students), Hradec Kralove (Czech Republic), Representative of the Faculty of Medicine Carl Gustav Carus (Dresden, Germany)*
Talk: 'Functional characterization of the human-specific gene *FAM72D,* and its ancestral paralogue *FAM72A*, in neural progenitor cells during development of the cerebral neocortex'